THE
TELECOMS COAST

A History of Terrestrial, Subsea and Space Communication in Cornwall

GEOFF VARRALL

First published in 2026 by Whittles Publishing, an imprint of Porto Press.

All rights reserved. No part of this publication may be reproduced, stored in a retrieval system, or transmitted in any form or by any means, electronic, mechanical, photocopying, recording or otherwise, without prior permission in writing from the publisher.

The editors have made every effort to ensure the accuracy of information contained in this publication, but assume no responsibility for any errors, inaccuracies, inconsistencies and omissions. Likewise, every effort has been made to contact copyright holders. If any copyright material has been reproduced unwittingly and without permission the Publisher will gladly receive information enabling them to rectify any error or omission in subsequent editions.

Copyright © 2026 Geoff Varrall

British Library Cataloguing in Publication Data
A CIP record for this book is available from the British Library

ISBN: 978-1-84995-611-6

The rights of Geoff Varrall to be identified as the author of this work have been asserted by him in accordance with the Copyright, Design and Patents Act 1988.

Cover and text design by Raspberry Creative Type, Edinburgh

Printed and bound in Great Britain by CPI

To order please go to our website www.portopress.com or contact our distributor, BookSource, 50 Cambuslang Road, Clydesmill Industrial Estate, Glasgow G32 8NB. Telephone 0141 642 9192

Porto Press Ltd
3 Connaught Road
St Albans
AL3 5RX

www.portopress.com

Paper from responsible sources

CONTENTS

Acknowledgements		v
Introduction		xi
1	Tin to telecoms	1
2	Steam to smartphones	15
3	Broad gauge to broadband	30
4	Subsea cable: Porthcurno as a portal to the rest of the world	44
5	Cable & Wireless	68
6	Steam radio to beam radio	87
7	The space, satellite and data centre story	112
8	The Telecoms Coast security story	124
Appendix 1: Geevor engineering and energy requirements		134
Appendix 2: Optical fibre technology		136
Appendix 3: Cables and batteries, link budgets and network optimisation		138
Appendix 4: Valentia Island Cable Station		143
Appendix 5: Marconi Archive, Bodleian Library: Poldhu		148
Appendix 6: Goonhilly Earth Station construction, and radio engineering documents		154
Bibliography and resources		158
About the author		161
Endnotes		162
Index		167

ACKNOWLEDGEMENTS

There are many people who have been helpful with this project, and many books that I recommend reading for additional background.

John Moyle's book, *Cornwall's Communications – with Special Places to Visit*, first published by Twelveheads Press in Truro in 2009 and updated in 2015, is concise but full of useful information.

John starts with the evidence of prehistoric beacons extending westwards across Britain from Carn Brea near Land's End and then references the six straw-and-pitch beacons used on 30 July 1588 to signal the progress of the Spanish Armada from Falmouth to Plymouth, a distance of 60 miles. It took just ten to twenty minutes for the beacons to get the message to Plymouth instead of a six-hour horse ride.

The semaphore started to take over from smoke from the Napoleonic era onwards. In 1814, the Isles of Scilly became connected to the mainland by a telegraph tower 165 feet above sea level as and when it wasn't rainy, foggy or dark.

In 1688 Falmouth was the main port for overseas mail, establishing the westerly packet routes to Nova Scotia, New York, the Caribbean, the west coast of Latin America and the southern routes down to Lisbon and Gibraltar and then to the East through the Mediterranean to Malta and the Middle East, the same routes that the cable to India would follow in 1870. Steamships were used from the 1840s.

By 1867 Truro and Penzance were connected by telegraph to all other major cities. As we shall see in Chapter 3, the railways helped to build the telegraph system, initially for signalling and then to transmit business and personal telegrams between telegram offices normally at or close to each railway station. The terrestrial telegraph system was nationalised in 1870.

In 1872 a reliable telegraph arrived in Falmouth, and the Eastern Telegraph Company was formed, the genesis 60 years later for the Cable & Wireless company.

John summarises the deployment of subsea cables, starting with the Indian cable via Portugal in 1870 to the last of the traditional telegram cables from Porthcurno to Harbour Grace in Newfoundland in 1952, through to an optical fibre cable to Kilmore Quay in Ireland in 2000. Early systems used William Thomson's mirror galvanometer and the siphon recorder. Performance improvements over the next 100 years were determined by improvements in cable quality and maintenance. From 1925 automatic regeneration, replacing telegraph operators at repeater stations, reduced costs.

There are some nice pictures of the bunker at Porthcurno (more about this in Chapters 4 and 5) and the telegraph hut on Porthcurno Beach, with a handy list of cable destinations (Gibraltar, Lisbon, Vigo in Spain, Madeira, Faial in the Azores, Newfoundland, Ireland and Bilbao). The cable story is completed with a short history of the training college at Porthcurno between 1870 and 1993 and the opening of the Museum on 28 March 1997.

John has generously shared with us his research on the history of the cable industry, including information from the Porthcurno archive on the maintenance costs of cable, covered in more detail in Chapter 4.

John's radio story starts with Michael Faraday (1791–1867), Heinrich Hertz (1857–1894), Édoard Branley (1844–1874, the man who designed the first coherer) and Oliver Lodge (1851–1940), who was noted for his work on the principles of resonance (also known as syntony) and was generally irritated when Marconi patented the idea in 1901 (British patent 7777, known as the Four Sevens Patent).

Part of the reason why Marconi set up a radio station in Poldhu 10 miles down the coast from his Lizard Radio Station was to test whether a radio receiver at the Lizard could withstand large amounts of transmitted radio signal energy from Poldhu. Tuning the aerial circuits to different resonant frequencies helped. There is a summary of how Ambrose Fleming designed the 25-kilowatt transmitter at Poldhu (the first use of condensers as a high-voltage energy store for the spark

ACKNOWLEDGEMENTS

transmitter circuits), the configuration of the aerial masts and the afterlife of the Poldhu site (longwave news bulletins to ocean liner passengers, and longwave telegrams). The site closed in 1933. Shortwave beam radio was also developed at Poldhu and is covered in Chapter 6. Beam radio, as the name implies, uses directional antennas. These had evolved largely by a process of trial and error, initially with the longwave array at Poldhu. John documents the use of longwave directional antennas for direction finding in the First World War with sites at Mullion (close to Poldhu) and Prawle Point in south Devon; the system was used to direct airships based at RNAS Mullion on anti-submarine sorties. John also covers the early history of Goonhilly, covered in more detail in Chapter 7.

David Barlow* at the Marconi Centre in Poldhu brought together the story of Marconi in Cornwall and Goonhilly in his book *From Spark to Satellite, Marconi in Cornwall*. David also produced a history of the Lizard Wireless Telegraph Station. This was the radio station used by Marconi to receive a 'long-distance' (over-the-horizon) transmission on 23 January 1901 from St Catherine's Point on the Isle of Wight. A few years later, in the early hours of Monday 18 April 1910, the radio operator on an American transport ship, SS *Minnehaha*, sent an SOS message after grounding in deep fog off the island of Bryher in the Scillies. The St Mary's lifeboat was sent to rescue the crew and passengers. This was the first SOS message handled by a coast station. David was also an expert on Captain Henry Round (more about Round in Chapter 6).

Also volunteering at the Marconi Centre, Robin Ridge has been helpful in providing information on the satellite technology heritage of Goonhilly. Terry Giles, Cliff Malcolm and Neil Whitham have also been generous with their time and advice.

Nigel Wall, who worked on Aerial 3 at Goonhilly, and was one of the team working for W.J. Bray on the Madley Earth Station (where the GPO (General Post Office) satellite circuits were transferred in 2008) unearthed engineering and specification documents which proved to be a treasure trove of useful information.

Alan Renton, when curator at Trinity House, produced a resounding work on foghorns called *Lost Sounds* (Whittles Publishing, 2001, reprinted in 2014). Any book about coastal communication and the closely related topic of safety of life at sea (SOLAS) could and probably should start with foghorns (audio communication) and lighthouses (optical communication.) Alan's book takes us from the clockwork-driven bells introduced in the early 19th century to reed foghorns powered by hot air engines in the 1860s to the new siren fog signal introduced into the pair of lighthouses on Lizard Point in 1878.

As Scientific Adviser to Trinity House from 1836 to 1865, Michael Faraday had been advising on the use of electricity as a power source, with dynamos driven by either steam engines or caloric engines (engines that used the expansion of air rather than steam to turn heat into mechanical energy, typically via a piston driving a flywheel). The advantage of a caloric engine was that it was faster to start than a steam engine – fog banks can roll in quickly – though, as we shall see in Appendix 5, caloric engines and oil engines, sometime known as hot bulb engines (the immediate precursor of diesel engines), were initially less reliable than steam.

Whether it is a foghorn or a subsea cable or radio or transport link, the ability to produce and use energy efficiently is a fundamental start point for any system designed 'to communicate at a distance' (a telecommunication system), a topic revisited in Chapters 2 to 7.

Getting a foghorn to be audible more than 2 miles from the coast was always tricky – but important, given that the lighthouse would be hard to see through a fog bank, particularly during the day. As Scientific Adviser to Trinity House between 1896 and 1911, John William Strutt (later Lord Rayleigh) experimented with a range of trumpet designs tested under a wide range of atmospheric conditions including wind, rain, fog, hail and sleet. His Theory of Sound, published in 1877 and 1878, set out the principles of half-wave resonance and the refraction and reflection of sound energy. He also worked out why there were silent areas and aerial echoes at sea: fog banks reflect acoustic energy. More importantly for our story, his principles of fading, which came to be known as Rayleigh fading, can also be applied (and are applied today) to radio waves, and his theories of wavelength mixing and cancellation and resonance are equally applicable to radio and optical system design. Usefully for Marconi, Rayleigh became an enthusiastic promoter of 'wireless electrical fog signals'.

ACKNOWLEDGEMENTS

If it happens to be foggy on your walk along the Telecoms Coast you might hear a modest 800 Hz peep from an electric emitter. Victims of satellite and coastal-based radio navigation, the foghorns of England (and Scotland, Wales and Ireland) have been decommissioned.

This brings us to our final acknowledgement, which is a thank you to Keith Whittles for commissioning *The Telecoms Coast*, and to the editing and production team at Porto Press, who have worked hard to produce a book which (to my eyes at least) looks handsome.

*David Barlow sadly passed away on 23 February 2025.

INTRODUCTION

Terrestrial, subsea and space communication in Cornwall

The Telecoms Coast tells the story of telecommunications in Cornwall from 1830 through to the present day, spanning the earliest days of the terrestrial telegraph network, the laying of subsea cables from 1850 onwards and the introduction of high-power long-distance radio from 1900 and satellite communication from the early 1960s.

For reasons that I hope will become clear, we start at Geevor Tin Mine in Pendeen, then head down the coast to Land's End, the most westerly point of the English mainland and the shortest route to New York. From Land's End we come first to Porthcurno, where the first subsea cable to India came ashore on 8 June 1870, then via Newlyn, Penzance, Marazion and Porthleven to Poldhu Cove near Mullion, from where (31 years later on 12 December 1901) Ambrose Fleming summoned sufficient radio energy to send the Morse code letter S (three dots, no dashes) to Marconi in Newfoundland.

From Poldhu we head a few miles inland to Goonhilly, where on 11 July 1962 (another 61 years on), TV signals from the newly built satellite communications antenna Arthur were transmitted and received via a beach-ball-sized satellite called Telstar. Fast forward another 60 years (2022 onwards) and those legacy dishes support NASA and ESA space missions. There is also a state-of-the-art data centre and a low earth orbit (LEO) ground station.[1]

A pilgrimage from Pendeen via Land's End to Porthcurno, Penzance, Poldhu, Goonhilly and Lizard Point

If you are a walker heading anticlockwise along the South West Coast Path, the distance from Pendeen (Geevor postcode TR19 7EW) to Land's End (the Land's End Hotel TR10 7AA) is 10.4 miles (16.7 km). From Land's End to Porthcurno (PK Porthcurno TR19 6JX) is 5 miles (8.1 km). From Porthcurno to Poldhu (Marconi Centre TR12 7JB) via Penzance and Porthleven is 32.6 miles (52.4 km). From Poldhu to Goonhilly Earth Station (TR12 6LQ) is 5.4 miles (8.64 km) and from Goonhilly to Lizard Point (the Housel Bay Hotel TR12 7PG) is 8 miles (12.8 km); total distance, 52.4 miles (83.84 km). To complete the tour, we could head north to GCHQ in Bude. Google tells us this is a 148-mile journey from end to end.

The Porthcurno Handbook, self-published by J.E. Packer in 1973, describes the beach at Porthcurno as 'made of sand composed of small shell fragments, white in colour giving the sea a rich emerald or turquoise hue'. Poldhu Cove, the other side of Mount's Bay, is equally lush. The swimming (at Porthcurno) and surfing (at Poldhu) is good as well.

Packer talks of Land's End as being 'tamed by tarmac', and urges visitors to head out from their deckchairs in the car park to the rugged coast path and the hinterland of 'wild crofts and moors dotted with the ruins of a decadent industry and stray vestiges in stone and tumuli left by a long vanished people to enjoy the seductive wildness and grandeur of the landscape to the north and west of Penzance'.

The Telecoms Coast is not written as a guidebook or gazetteer, but if you were minded to head down the coast from Pendeen with a compass and a smartphone you could be relaxing a mere 100,000 steps later with a well-earned gin and tonic in the bar of the Housel Bay Hotel. There, you can reflect that with the benefits of a time machine you could have crossed paths with John Pender in 1870 in Porthcurno as he planned the next long-distance subsea cable to arrive on the sandy beach. Fast forward 30 years to 1900, and you could have been chatting to Marconi and to Major Flood Page, the new director of the Marconi Telegraph Company, as they planned the new generation of high-power long-distance radio stations along the Cornish coast. A quick teleport

INTRODUCTION

to the present day, and you could find a bevy of satellite engineers from Goonhilly enjoying a beer or three in the hotel bar while discussing radio networks on the Moon or how to talk to Mars.

Cornish mining, Cornish steam power, steamships and the railway helped kick-start a long-distance communications revolution via Land's End and the Lizard – the Telecoms Coast – in Cornwall. This book helps to tell that story.

A note about coasts

The Tin Coast[2]

The Tin Coast is designated by the National Trust as a 7-mile stretch of the South West Coast Path, starting at Pendeen and heading down to Cape Cornwall, including the valleys to the south and three National Trust Properties: Botallack, Cape Cornwall, and the Levant Mine and Beam Engine, 'the best concentration of combined arsenic and tin processing sites in the world, and the only working steam-powered beam engine still in its original site'.[3]

The Tin Coast Partnership

The Tin Coast Partnership is 'made up of local businesses, charities and community groups who work together to develop sustainable and responsible tourism that will benefit those who live here, those who visit and the local economy'. The partnership includes Geevor Tin Mine, St Just Town Council and Cornwall Council. It also included Visit Cornwall until its demise in October 2025.

The Telecoms Coast

Starting at Pendeen (specifically the Geevor Tin Mine) helps us to tell the story of the technical and commercial touch points between the Tin Coast and the Telecoms Coast. From the Housel Bay Hotel it is a short walk to Marconi's wireless station and the Lizard Lighthouse, both owned and managed by the National Trust.[4, 5]

The Coal Coast[6]

Tin mining in Cornwall and coal mining in South Wales were closely interdependent. Coal from South Wales fed the pump engines of the tin mines along the Tin Coast. Tin was shipped back to Wales to be smelted as part of a two-way trade, via Newport, Cardiff, Penarth, Barry, Porthcawl, Port Talbot, Llanelli, Saundersfoot, Tenby and Milford Haven.

From 1900 onwards (with the Cornwall Electric Power Act of 1902), privately owned coal-fired power stations started to deliver electricity to the tin mines along the Tin Coast. As we shall see in Chapter 2, it took a while to connect the Telecoms Coast to the mains electricity supply, but they got there in the end.

The Cornish coast[7]

Cornwall has the longest coastline (321 miles/513 km) of any county in England and Wales, and 40 harbours including estuaries supporting inland harbours (Falmouth, the Helford River (Gweek) and Hayle). It is therefore a naturally well-connected county for a maritime economy.

The Fish Coast

Local records document a harvest of 122 million pilchards (now renamed 'Cornish sardines') along the Cornish coast during 1847. In May 1905 so many pilchards were landed at Newlyn that 50,000 fish had to be thrown back into the sea; the pilchard shoal had been reported as extending for a distance of 100 miles. In Chapter 2, I talk about the energy economics of the mining and telecoms industry in Cornwall. While fishing is comparatively less energy-intensive, when a catch is only marginally economic, the cost of diesel, even when taxed at a low rate, becomes material, as do refrigeration and transport costs.

A note about the structure of the chapters in this book

The working title of this book was originally *Cornish Connections*. Although the title has changed (mainly because the book is about places on or near the Cornish coast), the book is still organised to look at linear and lateral connections.

Linear connections are determined by dates. For example, Chapter 7 looks at how William Bray (John Bray) in the mid to late 20th century followed in the footsteps of William Preece in the mid to late 19th century. They were chief engineers in the Post Office and they had a profound impact on telecommunications, both in their own time and subsequently.

Lateral connections are connections between different industries or different people working related engineering disciplines. For example, Chapter 1 documents the ways in which the tin-mining industry shaped the Cornish economy, particularly in the 18th, 19th and 20th centuries, and covers the related links between tin mining and telecoms. You might find this leads to some surprising jumps between topics, but bear with us; it will all make sense by the time you get to the end of Chapter 8.

And an apology

It will also become obvious that the story of the Telecoms Coast revolves around a small cast of men. This can be explained, though not excused, on the basis that engineering records and documents have mostly been written by men, and engineering stories mostly written by men about other men. All that can be said at this point is 'must try harder next time'. An essay by Veronica Davis Perkins, 'Whose Line Is It Anyway? Women, Opportunity and Change 1830–1920', in *Semaphore to Short Waves*, published by the RSA in 1998, redresses this shortcoming to an extent. In the context of this book, Sophia Kingdom Brunel, Emma Pender and Annie Jameson helped to shape the Telecoms Coast, and their story deserves to be told in more detail.

The other point to make is that in terms of the cast of actors we should be talking Ben-Hur rather than Hamlet; technology and engineering progress is the product of thousands of scientists and engineers working towards a common purpose – not a few men and women treading the boards alone.

On the other hand, I am trying to produce a book small and light enough to fit into a backpack after a visit to Geevor, Porthcurno, Poldhu or the Lizard, rather than *War and Peace*. Life is full of ultimately sensible compromises, and writing to an agreed word count and page count is one of them.

1
TIN TO TELECOMS

From magma to mobile phones

Around 280 million years ago, superheated molten magma from below the Earth's crust began pushing upwards, beginning the process of breaking apart Gondwanaland to create what would become South America, Africa, India, Madagascar, Australia and New Zealand.

The heat and force of this intrusive process helped to create the gneiss granite outcrops that are the dominant feature of the Cornish coast but also started the process of mineralisation and deposition of metallic ores which would become the bedrock of the Cornish economy and lay the foundations of the Cornish tin- and copper-mining industry. The same processes produced the raw materials that enabled the deployment of subsea cable, long-distance wireless and satellite communication – the three complementary telecom systems that turned the Telecoms Coast into a powerhouse of global connectivity.

Like Gondwanaland, Cornwall was not called Cornwall until someone decided to call it Cornwall, its name derives from the Celtic tribal name 'Cornovii', meaning 'headland', or more generally 'the peninsula people of the headland'. In the late 9th century, the Anglo-Saxons added the plural of the old English word '*walh*', meaning 'stranger' or 'foreigner', on the basis that it took several days to get to Cornwall, and when you did you were not always welcome. In Medieval

Latin texts it is described as Cornubia. The Anglo-Saxons also needed a taxation system for the post-Roman era, which produced the concept of shires for the purpose of counting, hence the word (and title) 'count' and hence the word 'county'.

Mining started to be economically useful to Cornwall over 3,000 years ago. The mass of granite is known as the Cornubian batholith. The clay mineral kaolin proved to be one of the more useful deposits originally for the drinking of Indian tea in china cups. Other useful deposits, in addition to tin and copper, are lead, zinc, silver, gold, arsenic and iron ore. Initially, the mining of the metallic ores was a relatively simple process of sifting alluvial deposits produced from weathering and erosion, later supplemented by open-cast mining. From the Middle Ages onwards, and possibly earlier, Cornish miners started to dig tunnels, including from the cliff edges out under the sea.

Several thousand years ago, someone had the idea of melting flecks of tin ore over a fire (tin melts at a relatively low 232°C). Flecks of copper need a much hotter fire (copper melts at 1,083°C), and its extraction means that someone must have worked out that pumping air into a flame made it hotter. Charcoal and coal added yet more fuel to the flames. Then some other bright spark worked out that if you added tin to a vat of hot copper, the strength and hardness of tin combined with the malleability of copper produced an alloy that became known as bronze; the word probably derives from the Venetian word '*bronza*', meaning 'glowing coals', and/or the German word '*brunst*', meaning 'fire'. This marked the beginning of the Bronze Age in Britain (2300–700 BCE), which is generally accepted as having started the Bronze Age in Europe.

The Bronze Age was followed by the Iron Age (700 BCE–1 CE). Although there is evidence to suggest that the Egyptians produced good-quality iron 6,000 years ago from material extracted from meteorites, the mass-production of iron required centuries of experimentation and a gradual understanding of how the addition of carbon combined with the production process (wrought or cast) affected material hardness. In 1740 Benjamin Huntsman, a clockmaker and instrument maker in Doncaster, managed to produce crucible-cast steel that was sufficiently uniform and pure to be used in clock springs. In 1856, just over 100 years later, Bessemer's patent to mass-produce steel from pig iron extended the use of steel into every nook and cranny of modern industrial life.

Tin and telecoms, copper and communication

It might seem obvious, but the ability to mine metal and manufacture metal alloys was (and remains) fundamental to the telecommunications industry. Remembering the basic physics that we may have learnt but were certainly taught at school, the one thing that sets metals apart from minerals is that metals conduct and store electricity. (Some metals, such as gold and silver, are minerals *and* metals, but that is an additional detail.) Silver wins the conductivity prize, but tarnishes easily. Gold is resistant to tarnishing but is rarely used due to cost. As copper has higher conductivity than gold and is more plentiful and hence less costly (and does not tarnish easily) it has become the default option for transmission and interconnection in telecommunication and power networks. Adding zinc to copper produces a harder and more resilient material, which we know as brass (probably from the old German '*braznaz*', meaning 'brazen'). As the zinc ratio increases, conductivity decreases.

Arsenic, a by-product of many mining processes including tin, is also widely used in modern electronics, including gallium arsenide transistors and solar panels. Tin is used to solder electrical components and semiconductors onto circuit boards. Indium tin oxides are used in optical networks, because they are transparent to visible wavelengths/frequencies (from red at 650 nm/460THz (terahertz) to violet at 400 nm/750 THz). Usefully, they also absorb optical C band from 1560 to 1530 nm/191 to 195THz. (If you like technical detail, go to Appendix 2.)

Early landline terrestrial telegraphy mostly used iron wire, as it was less expensive than copper, less likely to be stolen, and adequate for the low transmission bandwidth and typical cabling distance between telegraph stations or repeater relays. But once telephony started to take over from telegraphy, copper or a mix of aluminium and copper became a necessary option.

The specification of materials and material quality therefore became progressively more important. This in turn translated into standards that could be applied to qualify and manage telephone cable and component manufacturers and suppliers. The *Telephone and Telegraph Engineers' Handbook*, published in 1908,[8] specifies three grades of iron wire: Extra Best Best (EBB) which has the highest conductivity, the most uniform quality, toughness and pliability; Best Best (BB) often passed off as Extra Best Best; and Best, the poorest grade, with less uniformity, more brittleness and lower conductivity. Other metrics include pliability (stretchability), ductility (twistability), tensile strength and weight per mile.

For subsea cable, copper was always the default choice, and today copper remains the only choice for modern power transmission cables, often now packaged with fibre cables within the same outer sheath, a staple product for the offshore windfarm industry. Copper is now beginning to be recycled in quantity. In the UK, British Telecom confirmed in September 2023 that it was halting the sale of copper-based telephone lines, and its installation partner, Open Reach, announced plans to recover 200,000 tonnes of copper through the 2030s over a 15-year period.[9] At present the value, including the cost of recovery and recycling, is of the order of £30,000 per tonne. This value could increase if recycling efficiency improves and the price of mined copper continues to go up.

Twisted pair and coaxial cabling

As signalling rates increased in terrestrial telegraph systems, interference and noise became an increasing problem. This was partially solved by the invention of twisted pair by Alexander Graham Bell in 1881. The need to support telephony (with 30 times more bandwidth than telegraphy) meant that twisted pair became ubiquitous; by 1900 the entire American telephone network was twisted pair.

In 1880 Oliver Heaviside patented coaxial cable, establishing the principle of an inner conductor surrounded by a concentric conducting shield. These cables had been used from the late 1850s to reduce signal interference between parallel cables, so the patent in effect described what already existed, which is not unusual in the patent industry. Long-distance coaxial cables were introduced in the United States in 1946 to support the rollout of analogue frequency division multiplexing, typically 1,800 two-way voice circuits supported by three cables. By 1978 this had increased to 13,200 voice signals for each pair of cables within a cable bundle of ten working pairs. Time-division digital coaxial circuits were introduced into the US long-distance network in 1962. By 1975 these systems could support over 40,000 voice circuits over ten working pairs of cable.

The introduction of digital signal processing gave copper networks a final lease of life. Twisted pair with an ADSL (Asymmetric Digital Subscriber Line) modem at either end of the link supported 1 to 2 Mbps over short distances (100 metres or so). Coaxial cables supported of the order of 200 Mbps but never replaced legacy copper twisted pair in what was known as the 'final mile' from the exchange to businesses and homes, due to the high sunk cost of the twisted pair network.

As long- and short-distance links scaled to hundreds of megabits per second, then terabits per second (with trunked links now supporting petabits per second), it became clear that the age of the copper cable for telecoms was coming to an end. The high maintenance cost of copper networks when compared to optical networks (which are more stable and generally do not mind getting wet) provided an additional economic incentive. Within the home, plastic optical fibre is replacing coaxial cable for higher bandwidth applications such as home cinemas.

Metals and batteries

Copper and lead, however, remain essential to telecom battery systems. The early terrestrial telegraph systems were powered by a mix of lead acid primary and secondary batteries. At time of writing there are still customers in the UK with fixed telephone lines powered by a 12-volt DC supply from the local switch. Batteries also powered subsea cables for the best part of 100 years (1870–1970), with, from the 1950s, the addition of a mains power supply along the copper sheath of the cable to power undersea repeaters (also used in subsea cables today).

Metals and capacitance

With the advent of long-distance longwave wireless from 1900 onwards, there was a need to produce high voltages to drive the spark transmitter circuits. Marconi recruited Ambrose Fleming because of his experience working for Thomas Edison on DC and AC power generation, with Poldhu as his first great high-power experiment (25 amps AC at 1500 volts for the primary arc transmitter transformed to 20 kilovolts to generate 13 kilowatts of RF power). These voltages required a massive amount of capacitance. At Poldhu, Fleming used 20 large glass plates coated on one side with tin foil suspended in a stoneware jug full of linseed oil. The technologies and techniques developed at Poldhu were then used by Marconi at Clifden on the west coast of Ireland, where an even larger 80 kilovolt transmitter was built in 1906; 1800 large galvanised sheets were hung vertically in a 100-metre condenser room, with the air between the sheets acting as a dielectric.

Metals and magnets

It would be hard to build a telecommunication network without magnets for transformers, detectors, dynamos and alternators. Likewise, magnets without iron.

Cornwall has a long history of mining iron ore. Going back to that lesson at school that we may have been awake or asleep in, magnets are either made from iron or cobalt, or from metal alloys which contain iron, for example steel. We need magnets to turn mechanical power into electrical energy. We need electrical energy for all telecommunication systems.

Copper and power transmission, and increasing copper prices

Modern wireless systems are many orders of magnitude more power-efficient and bandwidth-efficient than early spark transmitters, but a large cellular base station site still uses several kilowatts of AC power backed up by battery-stored DC. A modern hyperscale data centre uses upwards of 100 megawatts of power. The latest (as at 2025) AI data centres typically consume gigawatts of power. To put this into perspective, a large town consumes something of the order of 50 megawatts.

So, although the use of copper in telecommunications cables is decreasing, this is more than offset by the increasing need to move power around, and copper remains the best power transmission option. This demand is driving the increase in copper prices.

We may be using less copper in telecom cables, but there is plenty of high-value metal in a smartphone, and even though smartphone sales are reducing (we are keeping them for longer) we are still buying (and making) over 1 billion units per year. Looking at the Bill of Materials (BOM) of a phone, we can typically count over 60 types of metal, including rare earth metals used in the signal processing chain from the radio frequency (RF) front end to the capacitive touch screen display (indium tin oxide).[10]

Most of these metals are sourced from African mines, and/or mines in Australia, and/or South America (presently the world's major source of lithium). Political instability in Africa and South America is an additional factor determining the increase in copper prices.[11]

Apart from plans to extract lithium from geothermal waters in Cornwall, there is presently no metal or mineral mining in the county, but that may change if copper prices continue to rise and overseas sources become harder to access. As one example, at time of writing early financing was being put in place by a Canadian-owned company to restart tin mining at South Crofty, close to Redruth and Camborne, with £56 million pledged from the UK national wealth fund.[12] South Crofty was the last mine to close in Cornwall (in 1998) shortly after Geevor. The plan is to produce tin from 2028 onwards based on an estimated extraction cost of $14,500 per tonne and a market price of $35,000 per tonne.

Even if, to mix metaphors, the Cornish mining industry remains dead and buried, its legacy will live on. Over the next seven chapters I will argue that the telecoms industry is an important part of that legacy and will set out how and why the Telecoms Coast in Cornwall became the crucible for long-distance communications innovation and continues to this day to be a vital part of the Cornish and British economy.

Though it might seem tenuous to draw parallels between 3,000 years of Cornish mining history and 200 years of telecommunications in Cornwall, there are more lessons to learn and more pointers to the future than we might expect, and the mine at Geevor, close to the village of Pendeen, provides us with our start point.

Geevor and the Cornish mining industry

The Geevor Mine is part of a 48,700 acre area of Cornwall designated in 2006 as a UNESCO World Heritage Site, marking the period from 1700 to 1914 when the mines of Cornwall were producing the raw materials for Britain's industrial revolution, alongside technologies which would be critical to the development of modern communication systems. Between 1800 and 1850, the area produced two thirds of the world's supply of copper and half of the world's arsenic, and the best part of 2 million tons of tin – all of which needed to be transported to coastal harbours by road or rail. Mining skills were also in demand overseas, with up to half a million people leaving Cornwall between 1815 and 1920 to work in the gold and diamond mines of Africa, and in copper and tin mines around the world.

In the 1850s most of the initial financing for the first subsea cables from the Telecoms Coast came from John Pender's friends in the cotton industry (the story of Pender, the Cable King, is told in Chapter 4). The early cables were manufactured close to the River Thames in Greenwich, but it certainly helped that there were local skills available in Cornwall, a manufacturing supply chain for the mining industry which was capable of making the miscellaneous metal components used in telegraph systems, some decent harbours, an adequate road system, and a railway into Penzance which could get you from Cornwall to London in a day.

During the Second Boer War (also known as the South African War) in 1899–1902, a group of Cornish miners headed back home and took up a prospecting lease on the North Levant and Geevor, Wheal Carne mines. In 1896 tin had been £64 a ton, but by 1906 the price had increased to £182, prompting the St Just and Camborne mines to reopen. In 1906, the North Levant[13] and Geevor Company was registered, with West Australian Gold Fields as a shareholder. This company became Geevor Tin Mines Limited, with a share capital of £150,000. This is not dissimilar to the sums of money that Marconi was investing in ever more powerful radio systems at Poldhu (and Clifden on the west coast of Ireland, and Glace Bay in Newfoundland).

Through the first 85 years of the 20th century, the profits of the mining industry and the telecommunications industry waxed and waned in parallel with one another. Metal prices boomed in the First World War and so did telegram traffic; and Marconi's radio communication and direction-finding equipment was in high demand. The great crash in 1929 was a disaster for everyone and hastened the merger of Marconi's long-distance radio business with the subsea cable industry (the establishment of the company that came to be known as Cable & Wireless). The Second World War pushed up metal prices again, and increased demand for radio (and radar) systems. The Cold War from the 1950s onwards kept metal prices high, and the space race from 1958 onwards gave birth to the satellite industry. (The Marconi Company helped to build the first satellite dishes at Goonhilly.)

In 1985 the telecommunications industry started to build the first generation of cellular base stations. Forty years later as we walk along the Telecoms Coast, we are never far from a cellular radio mast, and we can continue to surf the Internet as we sail out to sea.

In October 1985 the international cartel agreement which had kept tin prices stable (otherwise known as artificially high) for 30 years collapsed in a sea of debt, and the world price of tin fell from £10,000 per tonne to £3,400 per tonne.

In April 1986 the Geevor Tin Mine closed 'in a blaze of publicity, protest and emotion'. Some workers were reemployed to remove the iron ore left in the stopes (the empty spaces created underground as part of the mining process), a tidy-up process that yielded 44,000 tonnes of ore, and a rescue programme in October 1987 brought the mine back into limited production. Without a sustained increase in the market price of tin, the decision was made in May 1991 to switch off the underground pumps and allow the mine workings to flood. But from this end point, let's head back to the start.

Tin routes and telecom routes

Mining exports from Cornwall are now considered to have started during the British Bronze Age (2400–800 BCE), about 3,600 years ago. This is based on the carbon dating of oak shovels and antler picks, and the discovery of artefacts in Germany containing Cornish gold (metals have unique geological 'fingerprints' that help to identify where they were mined).

Tin ingots from Cornwall (identified by the mix of tin and lead isotopes and trace elements) were found in a shipwreck in Israel, with the wreck dated between the 13th and 12th centuries BCE. Diodorus of Sicily, a Greek historian living in the 1st century BCE, writes about exports from Cornwall to Gaul from Ictis (modern-day St Michael's Mount) which then found their way to Germany and the Levant on overland trading routes.[14]

By 2000 BCE, spices from China were being traded along the maritime trade routes to the Middle East. Spices were lightweight and high value but were part of a mixed manifest which would have included silk, ivory, porcelain, metals and gemstones. By the time goods had been trans-shipped to Europe they had travelled the best part of 15,000 kilometres (9,330 miles /8,200 nautical miles).

Four thousand years later, Pender's first subsea cable from India set off on a similar journey across the Indian Ocean and into the Red Sea, and then across to the Mediterranean. The advantage of the subsea cable compared to terrestrial cable routes across Europe was that they largely avoided local country politics. Cecil Rhodes had similar issues when he tried to build an overland telegraph network from Cape Town to Cairo. Even with all the problems associated with laying a cable along thousands of miles of ocean seabed, subsea was simpler, hence the routeing into the Atlantic and along the coasts of Portugal and France to Porthcurno. The same constraints apply today; it is easier to take a fibre cable under the sea and around Africa than it is to interconnect 54 countries that do not trust or like one another.

There is no history of significant metal and mineral trading to the New World essentially until the 19th and 20th centuries, when arsenic became a valuable export as an insecticide for cotton crop dusting (to kill the infamous boll weevil), and for use on potato farms (to control the equally infamous Colorado beetle). Arsenic was also used in other products, including paint, and often made more money for the mines than the tin from which it was extracted.[15]

The discovery of gold in California in the early 1840s – the start of the Californian gold rush – also marked the first of several waves of skilled miners heading to America in search of what they hoped was an easy fortune. They were known as forty-niners, as the biggest exodus was in 1849. High-grade copper in Cornish mines was getting harder to recover (with lodes at 1,000 feet) and the competition from mines in America and Chile was bringing prices down. Tin prices, too, were being hit by competition from Bolivia. These overseas mines, however, needed blasting skills, timbering skills, ventilation skills and ore-processing skills. One of the first miners documented to reach the gold fields was James Rickard, who crossed the Panama isthmus into California with his stamp mill, the first to be put to work in the mines. Some estimates suggested that about 15 per cent of the skilled miners came from Cornwall (they were known locally as Cousin Jacks). There was a similar exodus to South Africa after diamonds were discovered in Kimberly in 1867 and gold in Witwatersrand in 1886.

So, what did tin mining do for telecoms? More specifically, what did tin mining do for telecoms in Cornwall – and even more specifically, what did tin mining do for telecoms along the Telecoms Coast?

I have made the point that communication networks could not exist without copper, and that copper was only mined once tin had been discovered; but in practice the amount of Cornish copper in today's cable, radio and space networks is insignificant. Most of the copper network buildout globally was in the 20th century, when Cornwall's share of global markets was in decline.

But as mining in Cornwall started 3,000 years ago, its impact needs to be judged over thousands rather than hundreds of years. It cannot have been a coincidence that a few years after Cornish tin started to be shipped to Gaul from St Michael's Mount, Julius Caesar invaded Britain (in 55 and 54 BCE). Over the next few decades, the Romans built a road system from London to a mile or two from Marazion, or what the Romans called 'the territory of the Dumnonii'. The Dumnonii were Iron Age Celts based in what is now Exeter. They had iron mines on Exmoor and oversaw getting tin into Europe. The Romans managed to avoid killing the Dumnonii and came to a satisfactory set of business arrangements which lasted for 400 years, at which point the Dumnonii took over again. In the meantime, the Romans had connected Cornwall, avoiding various boggy bits in Somerset along the way. Brunel followed a similar route with his railway (Chapter 3) and if you find any straight bits on today's A30 from London via Redruth to Land's End they are probably Roman as well. Caesar connected Cornwall, and he connected Cornwall because he wanted and needed Cornish tin.

From Redruth to Rio

Pre-Christian era mining is not unique to Cornwall. Just as Redruth gets its name from the red ore that used to flow down the river through the town (in old Cornish, Rydh Ruth means Red Stream or Red Ford), the modern mining megalith Rio Tinto Zinc is named after the Rio Tinto or red river mines in Huelva in southern Spain, where Tartessians, Iberians and the Romans mined tin as early as 3000 BCE. The Spanish government sold the mines in 1873 to a group of London bankers, who then founded the Rio Tinto Company. The company moved into Zambia in 1929. The Spanish mines were sold back to the Spanish government in 1954, and the company moved into South Africa, Canada and Australia and, in 2010, to Mongolia. For the past 25 years, Australian exports of precious metals to China have been a dominant revenue stream. A resurrection of copper and tin mining in Cornwall, if it happens, is therefore likely to be subsumed into what is now a thoroughly globalised supply chain.

Tin economics and telecom economics

At a basic level, mining for metal and minerals drives economic growth, and economic growth drives the need for connectivity, which means harbours, ships, roads and railways, and short-, medium- and long-distance communication.

The Romans connected Cornwall to London, but Cornwall was connected to the rest of the world via the harbours at what is now Plymouth, Falmouth and Charlestown to the south, and Hayle and the north coast harbours.

If we join the Telecoms Coast at the Lizard and head up the coast past Poldhu and Mullion, we arrive at Marazion (Ictis and St Michael's Mount) where those first shipments of tin started shipping to Gaul 2,000 years ago. We then pass Penzance, then Newlyn where the Pilgrim Fathers stopped in 1620 before heading to a new life in the New World. The next stopping point is Porthcurno, then Land's End, Botallack and Levant, and finally Geevor.

Geevor and South Crofty were two of the largest tin mines in Cornwall, and the last mines to close in the 1990s. Both mines had been keystones of the Cornish economy, along with several dozen smaller mines; tin was an industry that directly employed 25 per cent of the Cornish population. Although Porthcurno never employed more than a few hundred people locally, Poldhu a few dozen and Goonhilly a few hundred (adding together each of the companies on site today), these numbers hide a bigger story.

From tin mines to data mines

The global mining industry today has a turnover of over $2 trillion. The global telecommunications industry has a turnover of over $3 trillion. Whereas the profits of both industries are significantly determined by their energy efficiency, their market value is also determined by their environmental efficiency; both industries account for about 3 per cent of global energy consumption. In mining, 25 per cent of the energy goes into grinding rocks, and 45 per cent is accounted for by mainly diesel-driven machines to move those rocks around, with the balance going into ventilation and lighting and what used to be called Information Technology (IT).

In telecoms, energy consumption is split between transmission, storage and processing, with AI generating significant additional overheads. Market enthusiasts talk enthusiastically about data mining being the new black gold, and there is a grain of truth in their excitement. There is no point in mining tin unless we can get more money for the tin than it costs us to mine. There is no point in mining data if the energy cost exceeds the market value. Cryptocurrency is a parallel in which the value must exceed the production cost (the energy cost of trillions of cycles of computer processing).

It turns out that there are useful lessons to be learnt from 150 years of telecommunications deployment along the Telecoms Coast, and some surprising synergies with the mining industry, which can help us understand why the economics of the two industries are closely coupled.

A tin-mining and data-mining thought experiment

Isaac Newton (1642–1727) was fond of thought experiments; one of his favourites was based on firing a cannon ball at a high enough speed to ensure the cannon ball never came back to Earth. Effectively he invented the satellite industry 200 years before Arthur C. Clarke.

Our thought experiment is to stand by the famous signpost in the grounds of the Land's End Hotel and ask these questions:

- How much money will it cost to move a metric tonne of tin from here to New York, how long will it take and how much money will it be worth once it gets there?
- How much money will it cost to mine an exabyte of data (1,000 petabytes) from Goonhilly and move it to New York, how long will it take and how much money will it be worth when it gets there?

I aim to provide at least a partial answer to these questions over the next seven chapters.

2

STEAM TO SMARTPHONES

From moss to mobile phones

Around 300 million years ago, as trees were gaining the upper hand over primitive moss, the world entered an era of carbon capture. Trees rotted into swamps and were covered by sediment. Then tectonic forces – our start point in Chapter 1 – crushed, cooked and compacted rotting vegetation, producing peat and methane gas. Further compaction and heat turned the peat into lignite, then coal.

In a parallel process, generations of microscopic marine organisms, plankton and algae died, which means that 300 million years later we could build a fossil fuel economy, and that along the Tin Coast coal-powered steam engines could pump water out of ever-deeper mine shafts.

From deep mines to deep mind – gunpowder and Geevor

In 1680 Cornish tin miners started using gunpowder to blast their way underground; 'shooting the rocks' meant that excavating a foot of rock could now be achieved in a shift rather than a week. In 1831 William Bickford, a Devon man from Ashburton, invented the safety fuse, strands of rope wound round a central core of gunpowder. Apart from being a vital prop in Wild West cowboy films, this innovation made miners' lives less hazardous but still hard.

By the 1830s several mines had extended beyond 1,000 feet underground and thousands of feet out under the sea; Geevor Mine at its lowest point was 1,575 feet (480 metres) below the surface. These deeper mines needed to be constantly emptied of floodwater (a process known as 'dewatering'), the tin ore needed to be brought to the surface, and miners needed to be taken up and down on lifting systems (1,000 feet on a ladder two times a shift was time-consuming, taking up to a two hour journey each way, and dangerous).

Two men from Devon introduced steam to the mining industry. Thomas Savery (1650–1715) patented a pump (in 1698) with hand-operated valves to pump out water by using condensed steam to produce a vacuum. Thomas Newcomen (1664–1729) improved pumping efficiency by separating the condensing steam from the water by a piston, known generally as an 'atmospheric engine', as it was the pressure of the atmosphere on one side relative to the partial vacuum on the other side of the piston that made the piston move. Atmospheric engines started to be used in the Cornish tin industry from about 1710; they were reliable but used a lot of coal, which was expensive as it had to be shipped from Wales.

From 1775 Newcomen engines started to be replaced with engines designed by James Watt (1736–1819) and his partner Matthew Boulton (1728–1809). Watt's steam engines improved efficiency with the addition of a separate condenser. In 1712 Thomas Savery produced a book called *The Miner's Friend*, which estimated the number of horses needed to do the same amount of work as one of his machines. Watt decided that a horse lashed to a spoke turning a shaft was pushing with a force of about 180 pounds and could typically walk around a 24-foot circle 144 times an hour before needing another nosebag of oats. Dividing this amount of work energy by 60, he decided that a good, strong work horse was producing 33,000 foot-pounds of work in one minute, which would be sufficient to raise a 33-pound bucket of water from the bottom of a 1,000-foot well in 60 seconds; this he decreed to be equivalent to one horsepower. The problem with this is that there is no such thing as a standard horse. Comparing a pit pony to a shire horse or an Arab stallion is like comparing a Morris Minor to a Ferrari (a racehorse on

Watt's calculation would be producing over 12 horsepower). In 1882 the International Standards Community decided that one horsepower needed to have an electrical power equivalent, which they decided to call a watt, in honour of James, and the decision was made that a horsepower should be regarded as being equivalent to 745.699872 watts. A watt as a unit of electric power is defined as one joule of energy per second (James Prescott Joule 1818–1889). Remembering that physics lesson at school which is lodged somewhere in one of our temporal lobes, energy is the capacity to do work, while power is the rate at which energy is transferred, or work is done. Energy is measured in joules and calories; power is measured in watts.

Now, 250 years later, we still talk about the power output from engines in horsepower – but radio and optical designers think in watts and kilowatts, and the electricity grid is dimensioned in kilowatts and megawatts and gigawatts, and the time-based equivalents (kilowatt, megawatt and gigawatt hours).

In our pilgrimage from Pendeen to Porthcurno, our first stop-off point is the Levant Mine and Beam Engine.

The Levant Mine was first worked in the 1820s as a high-value copper mine. In 1840 Harvey's of Hale installed a double-acting steam beam engine with a 24-inch cylinder and a 4-foot stroke; but after the original engine had thrown its flywheel through the roof, it was rebuilt by Hocking and Loam in 1862 with a new cylinder and valve gear. The engine then worked continuously for 70 years, until Levant was abandoned in 1930. After being rescued by the Trevithick Society,[16] the engine is regularly 'in steam' (Tours need to be pre-booked via the National Trust.) Bear in mind this was one of six steam engines on the site needed for dewatering and to power the compressors needed to feed air 1½ miles under the sea.

By the 1860s there were at least 600 steam engines used in the Cornish mining industry, pulling water up from 2,000-foot-deep shafts, hoisting tin ore and miners to and from the surface, and crushing and sorting thousands of tons of rock.

A few minutes down the path to Botallack, and we are back in the tin business. In 1898 Botallack was barely breaking even but then had what was to prove to be a temporary reprieve thanks to a local boy who made good, Francis Oats (1848–1918). Having trained as a

Surface building remains including engine houses and chimney stacks at Levant Mine and Beam Engine. The footpath to and from Geevor can be seen at the upper right of the image, with the Pendeen Watch Lighthouse (built in 1900, designed by Thomas Matthews) in the background. (Copyright National Trust Images, David Noton, used with permission.)

miner at the School of Mines in Penzance, Francis headed for South Africa, and after various adventures became a director of de Beers Diamonds Consolidated and close friend of Cecil Rhodes, returning to St Just as a wealthy man.

Botallack closed for a major upgrade. The steam engines were replaced with four gas engines. The tin mine was reopened in 1906 and in 1911 electric pumps were installed, powered from the new power station in Hayle. The areas above ground (or 'above grass' in local mining speak) were lavish, with the electric generator being set on a floor of Italian mosaic; this was money spent to impress visiting dignitaries. Along the coast at Poldhu at the same time, Marconi was welcoming royalty to his newly minted high-power radio station. Looking successful was as important then as it is today, though it didn't help the Botallack Mine, which closed in 1914 having produced no marketable tin since reopening.

Francis had, however, bought a nice plot of land on Cape Cornwall, with views of the Atlantic and the Irish Sea, and between 1907 and 1910 he built a fine Arts and Crafts House at Porthledden. The house is still there and sold recently for £5 million.[17]

Oats made his money from diamonds and gold at a time when the realised price from the South African mines exceeded the mining, transport, distribution and trading costs. In Cornwall, however, the Tin Coast was occasionally profitable, but vulnerable to the cost of coal imported from South Wales.

In this chapter (and Chapter 3) I will show how the use of steam, electricity, diesel and hydroelectric power in the tin-mining industry in Cornwall established the principles of energy economics, which in turn determined the underlying economics of the subsea cable and high-power radio industry – and which continue to determine the delivery economics of today's subsea, terrestrial and satellite radio networks.

From moss to mobile phones via Faraday and Marconi

In his *Experimental Researches in Electricity*, published in three volumes between 1839 and 1855, Michael Faraday (Scientific Adviser to Trinity House at the time) documented his research on dynamos and electricity generation. In 1858 the first electric lighthouse was installed at South Foreland (Dover); it was powered by magnetos 8 feet in diameter with coils rotating around a fixed magnet. In 1872, the Lizard Lighthouse (actually two lighthouses side by side) had Siemens dynamos installed.[18]

The dynamos, working at 850 rpm, produced a 40-amp current at 50 volts, with six machines bolted to a bed plate. The machines were driven by a leather belt from an overhead shaft by three 10-horsepower steam engines, each with a single-acting vertical cylinder, with each engine adjacent to a brick-lined iron furnace and integral coke-feeding hopper. The 25-inch diameter piston had a 20-inch stroke and rocked a beam which drove a large flywheel at 60 rpm. The flywheel, 8 feet

6 inches in diameter, produced the downstroke, but also drove a small air pump and valve gear. The pump forced air through the coke furnace to create sufficient heat to expand the air to a pressure of 26 psi, an early example of a turbo-charged engine. In addition to powering the light, the compressed air powered the foghorn. In 1885 the Siemens dynamos were replaced by two new magnetos built by the Demeritens Magneto Electric Machine Company. Weighing 4½ tons, these machines produced 4½ kilowatts of alternating current at 830 rpm. These machines remained in service until June 1950 when the lighthouse was converted to work on mains electricity.

The lighthouse and its efficient (for the period) steam-driven generator is a short walk from the Housel Bay Hotel, and although I have not come across any specific documentary evidence, it seems more than likely that Ambrose Fleming, Richard Vyvyan or George Kemp and Marconi's engineering team would have made the effort to look at the installation, particularly as they were negotiating contracts with Trinity House at the time for the provision of radio systems for maritime safety.

The problem with coal was that it was expensive both to buy and to transport from Wales. By 1900 oil engines were becoming easier to install and run than a steam engine, so Ambrose Fleming chose a 32-horsepower Hornsby-Ackroyd oil engine driving a 25 kW Mather and Platt alternator as the power source for the Poldhu transmitter. These engines, which first appeared in 1890, were used in agricultural machinery, and by 1900 would have been reasonably reliable. (Though Appendix 5 has Fleming venting his frustration on their occasional moodiness and unwillingness to start.) The engines were relatively high-compression machines, similar in principle to the diesel engine. The oil would probably have been delivered in 42-gallon barrels (standardised in the 1870s).

From 1900 onwards mains electricity started to be used in the mines and for subsea cable, for maritime and terrestrial radio, and then from the late 1960s for satellite earth stations, with diesel engines as a supplementary supply. Without the mines (specifically Geevor) it would have taken longer for the electricity grid to get to the coast in Cornwall.

From Geevor to Goonhilly via Google

In the 21st century we are beginning to buy and use AI-powered smartphones; the data centres that answer the questions that we ask our phones use prodigious amounts of power for storage, processing and communication. Goonhilly is part of this new global network. To be profitable, Goonhilly needs to use power efficiently and have access to sources of energy that are economically efficient and environmentally sustainable.

Tin mines and telecommunication systems are both essentially transport systems. In a tin mine we move metal ore from underground for processing (smelting) and then onward shipment for, often, thousands of miles. In 19th-century telecommunication systems, we started with a few dots and dashes over terrestrial telegraph and subsea cable systems. In terms of today's metrology this is equivalent to a few bytes of information.

Distance is also an important metric. Irrespective of whether we are talking about subsea cable or terrestrial radio or satellite systems, the longer the link the higher the power needed to overcome transmission loss and signal distortion. However, there are limits to the amount of power we can use. In a cable this is determined by breakdown voltage. In a radio system it is determined by the level of interference to other users.

Telecommunication systems are measured by wavelength, frequency and throughput. The telegraph systems were measured in words per minute, using the word 'PARIS' as the basis for comparison, as it represented a typical word in terms of the number of dots and dashes needed. Analogue telephone systems were measured by the number of voice circuits that could be supported over one telephone line with a voice bandwidth (including compression) of the order of 3 kHz. Since the 1970s, as the telecommunications industry has gone digital (a 50-year transition), throughput has been measured in bits per second, and storage bandwidth has been measured in bytes.

The development and standardisation of the Internet in the 1990s produced standards for packet networks with standardised packet lengths. As the World Wide Web evolved in parallel with the Internet, a parallel set of standards evolved that defined how data is stored, processed and searched.

In the tin-mining industry, tin ore is mined and then shipped to a smelter, where the raw material, 'black tin', is processed to become 'white tin' (tin with the unwanted impurities removed, then melted into an ingot of a defined size, shape, weight and quality). The economics of the process are defined by the mining cost, including the energy cost, the processing cost and the transport cost. A tin lode is a finite resource; once it is mined, it cannot be replaced. Mines generally close at the point where it becomes uneconomic to mine, and that point is generally determined by the energy cost of mining versus the market value of the tin. The energy cost is the cost of digging, processing and transport. Storage is an additional cost and can be significant.

Back to tin mining and data mining

Telecommunications is similar, but with some differences. If we defined data mining as being the underlying value generator in a modern telecommunications network, then we are never going to run out of data – but, as I pointed out at the end of Chapter 1, if it costs us more to store and mine the data than the realised value of the data, then we no longer have a sustainable business.

In the 21st century, the new telecom titans (TikTok, Google, Meta, Amazon, Microsoft, Apple and all) store, process and move exabytes of data to help us interact digitally with the analogue world around us (an exabyte is 1,000,000,000,000,000,000 bytes; that is, 1 with 18 noughts behind it).

A hyperscale data centre can be defined as 'a facility hosting at least 5,000 servers in a building of at least 10,000 square feet (930 square metres) with at least 40 megawatts of capacity'. A typical Google site is well over 1 million square feet, with a power consumption of over 100 megawatts. As stated previously, AI data centres increase this power consumption to gigawatts.

There are presently two data hubs in Cornwall with the potential to become hyperscale data centres. Goonhilly has a data centre which could potentially be used to store the exabytes of data being generated from the thousands of satellites that are now observing the Earth at sub-metre resolution (Chapter 7). Bude hosts the GCHQ data centre, where Internet traffic is stored for lawful interception and analysis (Chapter 8). The ability to scale these assets is dependent on either their ability to produce the power on site or the National Grid being upgraded, or a combination of both.

The problem with upgrading the National Grid is that the process usually involves overhead pylons, which are generally unpopular, especially in Areas of Outstanding Natural Beauty (AONB). Although taking power cables underground is an option, the cost per kilometre can increase by as much as 15 to 20 times.

This is not a new problem. Cornwall has natural sources of power including water and wind, but onshore wind farms and solar farms have the same local planning challenges as overhead pylons. Windmills worked well for grinding corn for centuries, and water mills were used for grinding flour and corn and for mining, including in the china clay industry – but sources of onshore wind power, water power and solar power are not easily scalable to hundreds or thousands of megawatts.

The National Grid and other sources of power

The mining industry was an early adopter of locally generated power coupled to the grid as and when it became available and cost economic. The china clay industry in and around St Austell started to go electric in 1886. St Austell was the first Cornish town to have an electricity supply; in 1890 the St Austell and District Lighting and Power Company was formed, and in 1902 425 kilowatts of capacity came on stream at Carn Brea, powered by four Belliss and Morcom steam generators. In 1911, the Hayle Generating Station was commissioned, delivering coal-fired mains power as far as Penzance. Local power distribution would have been via power lines on telegraph poles (technically known as utility poles) – which of course is still, in many rural areas, the way that electricity is delivered on the final drop from the local substation. Geevor started taking power from Hayle in 1919 and joined what would become the National Grid in 1933 with diesel generators as back-up.

In 1936 the Cornwall Electric Power Act transferred ten local companies to the Cornwall Electric Power Company, which became the South Western Electricity Board in 1948.

The Central Electricity Board had been set up in 1926 to link (mainly coal-fuelled) power stations into a 132-kilovolt National Grid; this was joined by a 275-kilovolt super grid in 1953, delivered via overhead pylons. Serious research on solar power started in the 1950s (Edmond Becquerel had discovered the photovoltaic effect in 1839, so this had taken a while to get going) and Britain started building the first generation of nuclear power stations with the mission to replace coal-fired power stations, and with the aim of delivering 'electricity too cheap to meter'. The last coal-fired power station was closed in 2024, ending 142 years of steam-power generation. In 1991 the UK's first wind farm was built in Delabole in North Cornwall, with ten 400-kilowatt turbines, enough to power 3,000 homes, but not an AI-enabled hyperscale data centre.

The international grid, the Isles of Scilly and subsea power

In 1986 a power cable linked the UK and France, giving our National Grid access to French nuclear power.[19] In 1989 a 55-kilometre 33-kilovolt subsea cable connected the Isles of Scilly to the National Grid. The original diesel-powered power station on St Mary's had been commissioned in 1932; after the power cable was installed, the power station became the standby source of power, supplemented by six generators on the outlying islands. From St Mary's there are two 11-kilovolt subsea links to Tresco and St Martin's, with local feeds to St Agnes and Bryer. The system serves a population of just over 2,000 people (1,600 customers plus summer visitors). The Bryer and St Agnes diesel generators (200 kVa generators manufactured by F.G. Wilson) have recently been renewed. Local low-voltage distribution is via telegraph poles/utility poles from 63 substations with a mix of overhead and underground mounted transformers. The powerlines also support broadband based on the IEEE 1901 standard operating in the 2 to 30 MHz band using Orthogonal Frequency Division Multiplexing; this is similar to the ADSL modulation used over twisted pair, and not that different from the modulation schemes used in 4G and 5G cellular radio networks. The power and communication network is monitored using radio modems in the UHF band and 2.4 GHz Wi-Fi band.[20]

An interrupted power supply

This all worked swimmingly well until 2017, when a fault developed 17 kilometres from Land's End. The back-up power supplies cranked into life within two minutes, and the CS (Cable Ship) *Sovereign* was despatched from Portland to repair the link.

The company contracted to fix the fault (now known as the Global Marine Company)[21] traces its origins back to the 1860s. A remotely-operated underwater vehicle (ROV) was used to cut the cable and lift it to the surface for repair.[22]

In 2024 the world's longest land and subsea power cable, the Viking Link, came online, connecting the UK and Danish power grids via the UK's east coast. This 765-kilometre 525-kilovolt cable can supply 1,400 megawatts of renewable power to and from the UK and Denmark, enough to power 1.4 million homes.[23]

The war of the currents at sea

In the 1880s and 1890s, Thomas Edison, Nikola Tesla and George Westinghouse were battling over the future of the electricity system, with Edison championing DC (direct current, where the current flows steadily in one direction) and Tesla and Westinghouse promoting AC (alternating current, in which the flow of the current constantly alternates.) AC won the argument. Most submarine power cables are alternating current. Stepping up the line current reduces resistive losses.

However, as inverters have become more efficient, DC has become a potentially better option, particularly for longer-distance high-voltage links, where the DC lines can carry 1.4 times more power than AC lines of the same size (they are not limited by RMS voltage, and reactive losses are eliminated). AC subsea power cables generally have three separate conductors within the cable sheath. DC cables usually have a single-core conductor.

Sunshine and wind in Morocco

Whether DC or AC is used, all existing subsea power cable links could have been dwarfed by a plan to import renewable electricity from Morocco. There, the sun shines on most days and at sunset the warm coast draws a wind from the mountains, producing a combination of solar and wind power for 19 hours a day. A 4,000-kilometre power cable was proposed, starting in Morocco and ending at Alverdiscott on the north Devon coast via the coasts of Portugal and France (following a route along the French and Portuguese coasts similar to that of the 1870 Indian telegraph cable from Bombay to Porthcurno). The power output was calculated to be 11.5 GW coupled to 22.5 GWh/5 GW of battery storage. The proposed agreement with the National Grid was for two 1.8 GW links into the Devon interconnection point, enough to power 7 million homes or two AI hyperscale data centres.

This was not a venture for the financially faint-hearted. The project was expected to cost £22–£24 billion and at time of writing had not been given planning consent. Lobbying from the offshore wind power industry may have created additional headwinds for the project. Morocco is now planning to use sun and wind energy locally to host a new generation of AI data centres.

The logic of delivering power from a National Grid was originally determined by production economics; a large coal-fired or gas-fired or nuclear-powered power station was assumed to be more economically efficient than locally generated power – for example, as of today, diesel generator power supplies.

Using a data centre as an example, the problem with this is that for every 100 megawatts of server power we need 500 megawatts of input fuel to the power plant, taking into account generation losses, transport and distribution losses and cooling, lighting and back-up power requirements.

The environmental argument is that renewables change this economic equation, as wind power, solar power, wave power and water power are free at the point of production (though they all contribute to the world's slowing speed of rotation – the energy has to come from somewhere).

For the Telecoms Coast, planning is underway at the time of writing for a small forest of floating offshore wind turbines (FOWTs) capitalising on those strong winds blowing across the Irish (Celtic) sea. The UK government has a target to deploy 50 GW of offshore wind by 2030, of which at least 5 GW will be floating wind energy by 2035, with a further 12 GW by 2045.[24] There will be competing demands for this energy, including the need to supply charge points for electric cars at peak periods of the day.

The challenge is that this power then has to be distributed along the Cornish coast. In 2025, the National Grid announced a plan to install 6,000 electricity pylons as part of a $31 billion investment in transmission networks, but whether these will be heading to Goonhilly or Bude any time soon is open to debate. It is unlikely to win votes in the South West of England.[25]

Charles Rolls and Henry Royce to the rescue

Meanwhile Mr Rolls and Mr Royce, builders of Marconi's favourite motor car, have been busy coming up with another solution.

The company is suggesting that their new generation of small modular reactors (SMR) will be the most cost-efficient and environmentally efficient way to supply our energy needs, including the energy needs of power-hungry data centres in rural and deep rural locations.

These are nuclear fission reactors small enough to fit on a 2-hectare (5-acre) site, easily accommodated at Goonhilly or Bude, each producing a handy 470 megawatts with no pylons needed (or coal trucks, gas plants or diesel tanks). Rolls Royce could build two a year from 2032. Sizewell C in Somerset will produce about six times more energy (enough for 6 million homes), but it occupies a 175-hectare site at Hinkley Point and will probably take another 15 years to build. Note also that Cornwall does not have 6 million homes – even counting its second homeowners, who only use power for two weeks every year.

In February 2025 the UK government announced that the planning constraints limiting nuclear power to just eight designated sites would no longer apply. It is official; the Telecoms Coast can become a Nuclear Coast.[26]

Rolls Royce small modular reactors. (Image credit Rolls Royce Power Systems)

The reactors are also proposed as a power source for the Moon and Mars as and when (and if) we settle there; more of this in Chapter 7. Local planning restrictions will not apply, at least initially.

Note that at time of writing there are no planned deployments in Cornwall – initial potential sites are at Oldbury and Berkeley in Gloucestershire.

Last but not least, in February 2025 the French Atomic Energy Agency sustained a fusion reaction in its WEST Tokamak reactor for 20 minutes. That means a single gram of hydrogen isotope could replace 11 tonnes of coal, which would definitely spell the end of the coal age.[27]

Tin mining and the Telecoms Coast – evolving energy models[28,29,30]

Most, but not all, economists think that Cornwall will never mine metal and minerals again. Even if tin and copper deposits were economically recoverable, the local environmental impact would be unacceptable.

However, energy cost economics and environmental cost economics change over time. The big shift happening in mining around the world is that mines are replacing diesel power with electric power. The machines that burrow underground are becoming battery-driven and remotely controlled from the surface. This changes the economics of large mines but makes small mines more economic as well. For one thing, the mine of the future will have few people underground, and they will only be underground for essential maintenance. Sovereign supply-chain security may also mean that state subsidies are justifiable for marginally economic or loss-making mines. Steelmaking in the UK is a present example of strategic subsidy.

Machines that run on batteries are more powerful than their diesel-powered equivalents (around twice the horsepower and more torque for the same size machine) and batteries are becoming more efficient (a tripling of voltage over the past decade); an electric truck can carry 50 tonnes of ore versus 40 tonnes for a diesel truck, and is quiet and pollution-free. Similarly, an electric mining machine with the same power as a diesel machine will be smaller so can work smaller tin and copper lodes and can be recharged at the end of every shift. After 70 years, the diesel age is coming to an end in the mining industry.

Cornwall was the only county in England where metal mining was carried out without local fuel – and in the end that crippled the industry. It certainly meant that energy-hungry processes such as smelting (5 tons of coal for every ton of white tin) happened outside the county; Swansea became the greatest smelting plant in Europe due to its proximity to high-quality coal.

Data mining is the same. It makes sense to have hyperscale data centres close to where the work is done. For Cornwall that means Bude (intelligence mining at GCHQ) and Goonhilly (space data mining), but these are power-hungry beasts that need local power rather than power from distant places.

We revisit this topic in Chapters 7 and 8. For now, it is time to return to our time machine and make our way back to the steam age.

3

BROAD GAUGE TO BROADBAND

*How Brunel connected Cornwall:
how Cornwall connected the world*

From Paddington to Penzance

For those of us daft enough to still be living in London, the starting point for our pilgrimage to Pendeen, Porthcurno and Poldhu (and Geevor to Goonhilly) is Paddington Station, with a choice of daytime trains or the Night Riviera sleeper train to Penzance.

The train journey first became possible on 2 May 1859, with the opening of the Royal Albert Bridge across the River Tamar, between Plymouth and Saltash, into Cornwall. Designed to accommodate tall ships sailing in and out of Plymouth Sound, the bridge, as described in John Murray's contemporary *Handbook for Travellers* is an

> extraordinary viaduct, which for novelty and ingenuity of construction, stands unrivalled in the world, carrying the new railroad at a height of 100 feet above the water on 19 spans or arches of which two alone bridge the estuary in gigantic leaps of 455 feet by means of Mr Brunel's ingenious combination of arch, tubular girder and suspension chain.

Twenty thousand spectators turned up to watch the iron trusses being hoisted into place.[31]

Brunel was too ill to attend the opening ceremony but was taken over the bridge shortly afterwards in an open carriage drawn by a broad-gauge locomotive. His father had died ten years before but would have been impressed by what is still today one of the most iconic of the great Victorian bridges.

The three Brunels – tunnels, railways, bridges and viaducts, steamships and harbours

Marc Isambard Brunel (1769–1849), Isambard Kingdom Brunel (1806–1859) and Henry Marc Brunel (1842–1903) were all masters of the art of taking whatever materials were available and getting the best out of them. Henry, the second son of Isambard and his wife Mary Elizabeth, deserves more attention – he helped to design and build Tower Bridge, Penarth Harbour and Barry Docks on the Coal Coast, and he surveyed the first proposed route for the Channel Tunnel – but it is the first two Brunels who left the biggest legacy for the Telecoms Coast, so Henry gets the cold shoulder from us as well.

The name Isambard is derived from the Old German word '*Isan*', meaning 'bright iron', and '*bard*' meaning 'axe' or 'bright' (reflecting qualities of strength and brilliance). Marc was born in 1769, and studied hydrography, then spent six years as a junior officer in the French navy, which at least partially explains his son's interest in ships. Having returned to Paris in 1892, Marc had a narrow escape from the cutting edge of the steel guillotine (known grimly as the National Razor). Escaping to Rouen, he met a 17-year-old English girl, Sophia Kingdom. He travelled to America, built a canal, was appointed Chief Engineer of New York and built a cannon foundry. At a dinner party he met another Frenchman, who mentioned that the Royal Navy had a problem making rigging blocks for sailing ships and they needed 100,000 blocks per year. Marc designed a machine to automate the process. In 1799 he travelled to England, hoping to make his fortune, and married Sophia. He met Henry Maudslay, the founder of the machine tool industry and inventor of the screw-cutting lathe, the planing machine and the micrometer. In 1800 Marc patented the 'New and Useful Machine for Cutting One or More Mortices, Forming the Sides and Cutting the Pin Hole of the Shells of Blocks and for Turning and Boring the Shivers, and Fitting and Fixing the Coak Therein'.

Due to prevarication by the Royal Navy, however, this failed to make his fortune. Meanwhile, on 9 April 1806 Sophia gave birth to their third child, a boy, who Marc and Sophia christened Isambard Kingdom.

In 1809 Marc redesigned an army boot which was such a success that the Foreign Secretary, Lord Castlereagh, gave him a contract to supply the whole army. But the boot factory burnt down, and peace was declared, and by 1821 Marc and Sophia were in a debtors' prison. When the Tsar of Russia, Alexander I, offered to clear Marc's debts in return for a bridge over the Neva River, the Duke of Wellington intervened and arranged for the creditors to be paid off.

The tunnels

In the meantime, in 1807 an initial attempt had been made by Richard Trevithick (1771–1833) to design and build a tunnel beneath the Thames between Rotherhithe and Limehouse. Born near Redruth, Trevithick, at 6 feet 2 inches when fully grown, was a Cornish giant. Marc and Isambard Brunel, were comparatively short (about 5 feet 4 inches), quite useful if you are building tunnels.

Although Trevithick's final plan for the tunnel would have worked, his investors got cold feet. So, in March 1825 Marc Brunel started building the tunnel, using the concept of a travelling shield still used in contemporary tunnelling, but he fell ill and his young son, Isambard, was approved in 1827 as a resident engineer. A series of disasters happened, and Isambard nearly drowned in one of the floods caused by the roof collapsing into the tunnel – a case of wet feet and cold feet – but the tunnel was finally completed. The booklet, *The Brunels' Tunnel*, available from the Brunel Museum on the north side of the tunnel, tells this story in detail.

The railways

You might wonder at this point what this has to do with the Telecoms Coast, and the answer is quite a lot. Both Marc and Isambard became experts in tunnelling and the use of wood in tunnels, skills they shared with the Cornish mining community. This knowledge and experience was essential when Brunel came to build the tunnels and viaducts that levelled the gradients on the Great Western Railway route down to Bristol (the 1.8-mile Box Hill Tunnel in Wiltshire was the longest railway tunnel of its time)[32] and then on to Exeter and Plymouth – and then towards Falmouth, which was even hillier in places. Beyond Plymouth, several forests of timber were used on top of masonry piers capped with iron plates for the viaducts needed to cross the steep valleys between the high moors and hills in Cornwall.

The Great Western Railway, founded in 1833, was part of a plan to connect London to New York by steamship, competing with Liverpool (nearer to New York by sea, but further from London). The project received its enabling act of Parliament in August 1835. The first trains ran in 1838, with the initial route to Bristol completed by 1841. As Chief Engineer, Brunel managed every aspect of the design and engineering of the project (apart from the steam locomotives, see below).

Due to delays in upgrading the docks in Bristol (not Brunel's fault), Liverpool remained the main port of arrival and departure, but in the meantime, Brunel had designed some ground-breaking transatlantic steamships.

The steamships

Legend has it that at a GWR board meeting in Blackfriars in 1835, one of the board directors complained about the length of the railway line to Bristol (124 miles), missing the point that it would be difficult to move Bristol closer to London.

In January 1836 the Great Western Steamship Company was formed, with Brunel appointed as engineer, giving his services for free. The SS *Great Western*, launched in 1838, was the first steamship purpose-built for crossing the Atlantic and was used for transatlantic passenger travel between 1838 and 1846. The ship had iron straps round the wooden hull and was powered via side-wheel paddles. There were four masts to hoist auxiliary sails. At 212 feet (65 metres) long and a displacement of 2,300 tons, the ship was bigger and more efficient than any other ship afloat. As Brunel patiently explained to his shareholders (who were complaining about the inordinate size of the boat), 'the tonnage of a ship increases as the cube of its dimension while the resistance increases as the square'. The same principle applies to large rockets, which we will find is important in Chapter 7.

The first transatlantic sailings were advertised in March 1938, with 128 staterooms at 35 guineas each, and with 20 'good bed places' for servants, who travelled at half price. But there were teething troubles: on the initial sea trials, the boilers caught fire, and Brunel fell down a charred ladder into the boiler room. The ship was so large it could only get out of the dock without its paddles attached, so ended up working out of Liverpool.

The SS *Great Britain*, launched in 1843, was even bigger (322 feet/98 metres) and again was the largest and fastest ship at the time, built of metal and driven by a propeller rather than paddle wheels. Like the *Great Western*, the boat became stuck in the lock in Bristol, so also started service from Liverpool. By 1846 it was crossing the Atlantic on a regular basis, but ran aground off the Irish coast, possibly due to compass error caused by the iron hull. From 1847 it was used by the West India Mail Steam Packet Company (sailing from Falmouth), and it served as a troopship in the Crimean War. The ship was scuttled in the Falklands in 1937, then famously refloated in 1970 and returned to Bristol, where it was renovated and is now displayed. The point to this bit of the story is that the *Great Western* and the *Great Britain* were the most power-efficient and therefore cost-efficient boats of their time, and significantly helped to establish the Liverpool to New York transatlantic liner business. Fifty years later, Marconi was busy adding radio systems to the replacements for these ships, most famously the RMS *Titanic*, built on similar principles to Brunel's last, biggest and fastest steamship, the SS *Great Eastern*.

Originally known as the Leviathan the Great Eastern was launched in 1859. At 692 feet (211 metres) long and 120 feet (36.5 metres) wide it was also the most economic steamship in the world, capable of steaming to Australia, India, or America and back on one load of coal (18,000 tons in the double hull) with 4,000 passengers or 10,000 soldiers on board.[33]

This ship had teething troubles and more, too; one ship worker lost his life on the first – failed – attempt to launch the ship (sideways) from the Wapping Wharf (the remains of the wharf are still visible today). On the sea trials, an explosion in the forward funnel killed five stokers. Already ailing from a heart attack on board the ship before it was launched and with a host of other medical issues, Brunel died broken and disheartened on 15 September 1859.

The boat was never financially viable as a passenger liner (despite nine transatlantic trips including taking 2,000 troops to Quebec in 1861) but as we shall see in Chapter 4, the ship was uniquely well suited to laying Pender's cable from Bombay and the cable lays across the Atlantic.

But for the moment let us return to the railway and railway gauges.

Back to the railways

Narrow gauge

From the 18th century onwards, narrow-gauge railways had been built underground and above ground in tin and copper mines to connect the mines to the nearest or most convenient harbour.

Most of these railways were 2 feet/610 mm or narrower; the narrower gauges were generally used underground or within the mine above ground. Between 1911 and 1991 Geevor was running 18-inch (457 mm) track (similar to South Crofty). The Levant, a mine running out under the seabed for over a mile, had a 15-inch (381mm) track in use from before 1820 and still in place in 1930. Before steam, pit ponies and pack horses, mules and donkeys or men were used to move the trucks along the tracks (generally known as tramways).

Standard gauge

Robert Stephenson, the original Rocket Man (and a close friend of the Brunels – Robert was the only man that Isambard Kingdom considered his equal as an engineer), had decided in the 1830s that 4 feet 8½ inches (1,435 mm) was the best option for steam engines and steam trains; after all, it was more or less identical to the gauge used by the Romans (the distance between chariot wheels). Why reinvent the wheel if it had already been invented?

Broad gauge

Brunel, however, was having none of that, and decided that the ideal gauge was 7 feet (2,140 mm). The main reasons for this were high-speed stability and comfort, as the wider track allowed the trains' centre of gravity to be lowered.

Uncharacteristically, Brunel had underestimated the amount of power needed from the steam engines to take advantage of the high-speed capability of the broader gauge. In 1838 a 21-year-old engineer, Daniel Gooch, was brought in to do a redesign and to construct a purpose-built factory in Swindon.[34]

The result was the Iron Duke class of locomotives introduced in the late 1840s, which, equipped with 8-foot driving wheels, could reach 80 miles an hour. This was the fastest railway in the world. In terms of space, if you have spent the night in a cabin in the sleeper train to Penzance you will appreciate how much difference an extra 2 foot and 3 inches (705 mm) would have made. The drawback was the additional cost of buying land and building tunnels and bridges and stations to accommodate the wider tracks and trains. For those several reasons the Track Law Act of 1845 mandated Stephenson's 1,435 mm as the 'standard gauge'. This created a problem, because although the mainline down to and through Cornwall was Brunel's broad gauge, the newly built West Cornwall Railway was laid as standard gauge. In 1866 it was upgraded to mixed gauge, and on 1 March 1867 through trains from London to Penzance started running. They included fast services such as the 10.15 Cornishman and the 11.45 Flying Dutchman – but the journey still took at least nine hours.

Marconi and the standard gauge railway in Cornwall

In 1892 GWR carried out a broad gauge to standard gauge conversion. In July 1904 it introduced a new express service, leaving London at 10.10 and arriving at Plymouth 4 hours 25 minutes later, and Marconi, having taken this train, would have carried on to Gwinear Road station before heading down the branch line to Helston. There he kept a motorbike (essentially one of those bicycles you find in France with an engine over the front wheel, still used for pacing in keirin cycle racing) and would have used it for the final dash to Poldhu.[35]

Down the road at the Lizard, Marconi asked a local builder, George and Sons, to move the GWR waiting room in the village (for the omnibus link to Helston) to Bass Point, where it served as the

The Lizard Wireless Station, formerly the GWR waiting room (Copyright National Trust Images/David Sellman, with permission)

wireless hut (now owned and preserved by the National Trust). More about this in Chapter 6.

The Broccoli Coast

The Cornish branch lines lived on for a while through the 20th century, partly through summer tourism but also from the transport of fruit and vegetables and fish. By the 1930s the Helston line was surviving mainly on moving coal on the down line and broccoli on the up line (33,000 broccolis in 1936) plus some local granite. Competition from road transport was at that time marginal – but by the 1960s it was game over for many of the branch lines, including Helston, felled by the Beeching axe (though some of those lines, including Helston, have been resurrected as tourist attractions).

Why bother to go to Penzance?

A holiday in Cornwall was not as popular in 1860 as it is today. The Brunels unsurprisingly used the train to go on family holidays, but their favoured destination was Torquay. Bradshaw and his eponymous Railway Guide, popularised by Michael Portillo on his televised railway journeys, described Cornwall as 'One of the least inviting of the English Counties … a dreary waste'. By 1850 Cornwall had become as industrialised as the Midlands and the North of England, and unless you particularly enjoyed the smell of smokestacks and the whiff of arsenic it was an unlikely choice as a holiday destination.

Fortunately for the Great Western Railway there were other reasons to go to Penzance, and plenty of goods to transport in the other direction.

Two hundred years earlier (1663), Penzance had become a coinage town. Tin ore was brought to coinage towns to be assessed for the tin tax (from 1838 this tax went to the Duchy of Cornwall). As a result, Penzance became the economic centre of the mining business, with smelters setting up around the edges of the town. Mining continued to account for most of the money being made in Penzance up till the 1850s and helped to finance upgrades of the harbour, including a new lighthouse; in 1880 a dry dock and a marine foundry were set up by the Holman Foundry Company working on 700-ton ships. In 1801 the official census for Cornwall recorded a population of 192,281 people. By 1861 this had almost doubled, to 362,343.

From then onwards Cornwall's population started to reduce. Between 1871 and 1881 the population of the county fell by 9 per cent and the mining population by 24 per cent, many miners seeking their fortune in America or South Africa, or indeed anywhere where mines were opening and mining skills were in demand. During the 1860s no less than 9,000 people left Penzance, so there would have been more than a few miners taking the train to Truro on their way to more distant underground destinations.

The Cabbage Coast

Initially the other significant source of income in Cornwall, apart from fishing, was agriculture, though the potato blight in the 1840s and the repeal of the Corn Laws in 1846 had been problematic, and much of the county was effectively still subsistence farming.[36]

But then the arrival of Brunel's railway proved to be the gift that kept giving for what would now be described as the Cornish agricultural and fishing ecosystem. By 1862, three years after the through service to London had been established, 2,000 tons of fish and early potatoes per year were being sent from Penzance, and the intermediate stops along the line, to London and the industrial towns of the North. Early crop potatoes grown through the warm winter and spring in Cornwall were highly valued. By the end of the century 500 tons of daffodils, violets and anemones were heading out of the county each year, and over 3,000 tons of spring cabbage and broccoli. The diet of the people in the North was improved by Brunel's railway.

From 4 May 1859 the ever increasing flow of people and produce followed a railway timetable set on London time (until that date, 22½ minutes ahead of Penzance time). Synchronised time from Paddington to Bristol had been established in 1840, and now, 19 years later, Cornwall was directly connected to the modern age.

Cornwall invents the steam locomotive; steam connects Cornwall

It is however worth highlighting the contribution that Cornish engineers made to the development of the steam railway. The Birmingham-based power couple, Boulton and Watt, had dominated early steam technology, making complete engines rather than castings. James Watt's first steam engine with a separate condenser in 1765 marked a big leap in steam efficiency (three times more efficient than the Newcomen engine), but by 1800 the Boulton and Watt patents were expiring, creating opportunities for Cornish foundries to build their own high-compression high-efficiency steam engines.

Harvey's Foundry and the Cornish Copper Company, both based in Hayle, supplied most of the steam engines used in the Cornish mining industry. Richard Trevithick worked for Harvey's, and in 1800 he developed a high-pressure boiler that represented another big leap forward in steam efficiency. Good-quality steam coal from South Wales helped as well. In 1801 Trevithick demonstrated the first workable locomotive, and by 1809 a tramway linking Portreath to the copper mines around Redruth was up and running and 100 mules were heading for retirement (or possibly the local slaughterhouse). Harvey's and the Cornish Copper Company built steam boilers for steam trains (and steamships) for the rest of the century.

Steam trains stayed in service from Paddington to Penzance until the late 1950s. In 1964 diesel hydraulic engines were introduced,[37] maintaining a four-hour schedule to Plymouth with a 500-ton train and a stop at Taunton. At time of writing, the Night Riviera was still using 40-year-old HST (High Speed Train) locomotives, one at either end of the train. Each engine produces 2,500 horsepower.

Brunel and the A roads into Cornwall

Brunel's contribution to the road links into Cornwall and the Telecoms Coast is tenuous but worth noting. When the Romans left Britain, they left behind them a road system that centuries later would support a public stagecoach service which could average 96 kilometres per day.

Julius Caesar himself was recorded as being able to travel up to 160 kilometres a day, presumably with the occasional rest break. For the next 14 centuries not much happened to the road system, and the journey down to Cornwall would have become slower rather than faster.

The 1663 Turnpike Act, legalising road tolls in return for local road maintenance, started to make a difference, but it was John Macadam (1756–1834) who reinvented the Roman roadbuilding system, using materials of mixed particle size and a predetermined structure. Thomas Telford (1757–1834) also made a difference. Macadam, however, had a specific link to Cornwall, having taken up a career as a victualling officer in Falmouth in 1778. Not impressed by the state of Cornish roads, he wrote *Remarks on the Present System of Roadmaking* (1816) and *A Practical Essay on the Scientific Repair and Preservation of Roads* (1819), and in 1827 was appointed as the Surveyor of Roads. Cornish greenstone was, fortuitously, a good material for roadbuilding, and from 1890 onwards it was quarried in Newlyn. In 1901 Edgar Hooley, a surveyor working in Nottinghamshire, noticed that a barrel of tar had burst and mixed with a stretch of stone-built road, creating a remarkably smooth and durable road surface. This was the genesis of the composite material which he christened tarmacadam; the Tarmac Company was established in 1905.

The modern motorist heading to the South West from London today has two options: the M3/A303 or the M4/M5. While those roads have turned much of the A30 into the scenic route it's still there, wending its way through rural southern England, and 2025 marked the 100th anniversary of its arrival at Land's End. The road starts at Henly's roundabout, branching from the A4 near Hounslow, before passing the Southern Perimeter Road of Heathrow Airport. Six hours after leaving London (depending on holiday traffic) you will, if you continue on the A30 from Exeter, pass what remains of Marconi's shortwave beam radio station at Innis Downs on Bodmin Moor on your way down to Land's End. However, if you are heading for the Lizard you might, on passing Exeter, choose the A38 as an alternative (then the A390 to Truro and Penryn, the A394 to Helston and the A3083 down to Lizard Point).

As you leave Plymouth you cross the Tamar on the suspension bridge completed in 1961 using at least some of the techniques pioneered by Brunel. You will also have a fine view to the left of the Brunel railway bridge.

At a local level, in 1877 a trip to Penzance from Porthcurno would have involved hiring a horse and trap from Farmer Ellis for the grand sum of 10 shillings and 6 pence. By 1912 35 years later, holidaymakers were heading for the beach in motor cars, with the added attraction of being able to see the operators at work in the instrument room, an early example of technology tourism, although this was discontinued due to the 'risk of attacks by suffragettes'.

Brunel gets the telegraph down to Cornwall

But to finish back on the railway again, the one thing Brunel needed for his high-speed railway was a signalling system which could tell the train driver to stop when something had happened further down the line.

In 1837, four years after the Great Western Railway had been founded, William Cooke (1806–1879) and Charles Wheatstone (1802–1875) patented the Cooke and Wheatstone telegraph, using the principles of electromagnetism first demonstrated by the Danish physicist and chemist Hans Christian Orsted (1777–1851), who had discovered that an electric current in a wire deflected a compass needle.

It was an uneasy partnership, however. Cooke enjoyed inventing things and making money through patenting, but Wheatstone was an academic who liked to publish his results (and also liked making money). The two men ended up in court, with Marc Brunel acting for Cooke and John Frederic Daniell, an English chemist and physicist, acting for Wheatstone. The dispute was resolved when Cooke bought out Wheatstone's interests with a royalty agreement. The telegraph patent made significant sums of money for Cooke, which he managed to lose by investing in ill-fated mining ventures. Wheatstone did rather well from his royalties.

The patent was one of the last signed by William IV before his niece Victoria became queen. This was the beginning of railway mania in Britain; the end of the wars in France had meant the government no longer needed to borrow money, which meant that there was low-interest finance available for new business ventures.

Marc's (and Isambard's) support for Cooke was therefore more than altruistic. In July 1837, only a month after the grant of the patent, Cooke was demonstrating his telegraph system to the London and Birmingham Railway Company in a carriage shed in Camden, using 13 miles of five-wire circuit made of 70 miles of copper wire.

It was Robert Stephenson, the legendary Chief Engineer of the London and Birmingham Railway (and best buddy of the Brunels, see above), who had introduced Cooke to the Brunels. The GWR adopted the Cooke and Wheatstone system for signalling but also used the system for the first commercial telegraph service, which started on 9 April 1839 initially along 13 miles (21 km) of line between Paddington and West Drayton.

In 1863, three years after the regular timetable had been established between Paddington and Penzance via Bristol, Bradshaw's Guidebook records telegraph stations at Southall, West Drayton, Slough, Maidenhead, Twyford, Reading, Pangbourne, Didcot Junction, Steventon, Farringdon Road, Shrivenham, Swindon Junction, Wootton Bassett, Chippenham, Corsham, Bath, Keynsham, Bristol, Highbridge, Bridgwater, Taunton, Wellington, Cullompton, Hele, Exeter, Exmouth, Dawlish, Teignmouth, Totnes, Plymouth, Devonport, Lostwithiel, Truro, Redruth and Penzance.

Brunel's railway had brought the telegraph within striking distance of Porthcurno. Brunel had connected Cornwall. And Cornwall was now ready to connect the world. [38] [39]

4

SUBSEA CABLE: PORTHCURNO AS A PORTAL TO THE REST OF THE WORLD

John Pender (1816–1896) and Porthcurno

In 1852, as Brunel's Great Western Railway and its trackside telegraph edged steadily closer to the Cornish coast, John Pender, a 36-year-old Scottish-born textile merchant based in Manchester and London, was looking for somewhere to invest the fortune he had made from sourcing cotton in the USA and trading finished cotton cloth to India and China.

His new (second) wife, Emma Denison, suggested he should buy shares in the English (later British) and Irish Magnetic Telegraph Company (E&IM), based in Liverpool and Dublin. After successfully laying a cross-channel cable to France in 1851 (a distance of 21 miles from Dover to Calais), E&IM was given permission to lay an Anglo-Irish cable, which in turn would link with a transatlantic cable from Valentia on the west coast of Ireland to Newfoundland, a distance of 1,660 nautical miles. This venture was established as the Atlantic Telegraph Company with John Pender as one of 340 investors. Sharing the risk turned out to be well advised, as the cable never worked properly and never realised significant revenue.

In 1865 a second attempt at laying a cable was made, using the SS *Great Eastern* bought at auction for a fraction of its build cost.[40] This was a mammoth operation, involving a crew of 500, a cow (for milking), several pigs (for eating) and 12 oxen. The cable was lost two thirds of the way across, at a point where the ocean was 2 miles deep.

On the basis of third time lucky, Pender started a new company, the Anglo-American Telegraph Company, to take over the project, with each partner investing £250,000. No less than 9,000 tons of cable was loaded on to Brunel's great ship. The cable, newly designed by Sir William Thomson (later Lord Kelvin) was three times the diameter of the earlier cable. Within two weeks the cable was laid, and on 29 July 1866 Queen Victoria and United States President Andrew Johnson exchanged a congratulatory telegram. On the voyage home, the old cable was grappled from the seabed and repaired, resulting in two Atlantic cables for the price of one. This successful venture, overseen by the company's young Chief Engineer, Charles Tilston Bright (1832–1880),[41] marked the start of 30 years of cable laying, connecting Britain to Gibraltar, Malta, Bombay, Singapore, Australia, Brazil, Chile, Argentina and Africa, and a cable empire that was to last for 100 years; the last telegram cable was decommissioned in Porthcurno in 1970.

By the 1860s the introduction of screw-propeller steamships with iron hulls (thanks to Brunel) had reduced the transatlantic crossing time to 8 or 9 days. Even taking into account the need for message decoding and recoding, a telegram from London to New York could be sent in a few minutes. Today a message sent by vacuum-packed optical cable (light travels at 300 million metres per second in a vacuum) makes the journey in less than 60 milliseconds.

Back in the 1870s reducing days and weeks to minutes was sufficiently compelling to bring in a wide range of investors. Over the next 30 years, John Pender set up 30 telegraph companies. This approach meant that partners (including politically important country partners) could be brought into the venture, and if the venture failed other cable companies (and countries) would not be affected. Thirty years later Marconi adopted a similar approach.

In 1862 at the age of 42, John Pender became the Liberal Member of Parliament for Totnes, and in 1872 MP for Wick in Scotland. Whereas Totnes, one of the few remaining rotten boroughs, was undemanding,

Wick was more onerous due to troublesome disputes involving sheep – but Pender could at least defend the subsea cable business in Parliament as and when required. Politics and telecommunications have always been closely coupled.

The Atlantic cable route was the most successful and profitable cable route and helped to finance the merger of the Anglo-American Telegraph Company, the Gutta Percha Company and Glass Elliot (a cable manufacturer) to form the Telegraph Construction and Maintenance Company (Telcon). This company came to dominate the global cable manufacturing, laying and maintenance business, with Sir James Anderson at the helm as Managing Director. Sir James had been the highly respected and skilled captain of the *Great Eastern* and later became Managing Director of the Eastern Telegraph Company.

In 1868 Disraeli's Telegraph Act nationalised the terrestrial telegraph network; this was the first industry to be nationalised in Britain. The Post Office had to buy Pender's E&IM shares. Conveniently, this provided cash for a new venture, the Falmouth, Gibraltar and Malta Telegraph Company, set up to establish a cable link from Bombay. The cable was to be routed via the Kooria Mooria Islands in the Indian Ocean to Aden (now in Yemen), then along the Red Sea and through the Suez Canal (which had opened in November 1869), then Susa in Libya, then Malta and Gibraltar, then Carcavelos in Portugal and along the French Atlantic coast, terminating in Falmouth, which had the advantage of good existing land line connections to London.[42]

Falmouth and the Royal Mail Packet Service

Charles I (1600–1649) had started the King's Post with two mail routes to the Continent, one from Dover to Calais/Ostend and the other from Harwich to the Hook of Holland – but as soon as Britain started the war with Catholic France, the overland routes to the Mediterranean (Gibraltar, Malta and Corfu), Spain and Portugal (and on to America and the West Indies) became unusable. So in 1688 Falmouth was appointed as the Royal Mail Packet station, routeing messages and money to and from the expanding British Empire.

Although it was 273 miles from London to Falmouth down the Great West Road, parts of which were a rough track, Falmouth had the advantage of being relatively far away from the French coast. The packet vessels were typically two- or three-masted ten-gun brigs and often transported bullion, making them vulnerable to attack.

However, as steamships became fast and reliable (back to Brunel again), the ships could reach the safety of London quickly and in any weather; by 1850 the Falmouth packet era was more or less over. But as far as cables were concerned, its harbour was still considered too busy, with too high a risk of cable damage from dragging anchors, hence the decision to move along the coast to Porthcurno.

The cable from Bombay to Aden was laid by the *Great Eastern* (and four support vessels) with the lay from Malta back to Porthcurno starting on 14 May (1,150 nautical miles to Gibraltar, 366 nautical miles to Portugal and then 824 nautical miles to Porthcurno), arriving on 8 June 1870.

Pender did not have a monopoly of the cable business and neither did Porthcurno; the Western Union Cable from America came ashore at Sennen in 1881 and had offices in Penzance. In addition, overland routes through Europe provided alternative routes to the East. By 1872 the first telegraph system had reached the Lizard to connect Spain to Cornwall, with the cable coming ashore in Housel Bay.

Pragmatically, Porthcurno became a preferred landing point because it was a sheltered quiet cove with land available for building, well suited to connect with the maritime routes which the Royal Navy were already set up to patrol. The combination of the best and most secure undersea routeings to both America and India, with minimal routeing over land, meant that by 1896 (the year he died), Pender was wealthy and influential. You do not need a monopoly to dominate a market, but you do need technology that works and continues to work for long enough to make a profit from the investment.

The business of making subsea cable work

The point about the role of the *Great Eastern* establishing the early long cable lays has already been made alongside the role that William

Cooke and Charles Wheatstone played in establishing – and, in Wheatstone's case, perfecting – the early terrestrial telegraph; but it was Wheatstone who made the telegraph suitable for undersea use. He had been one of the first people to support the idea of underwater telegraphy in 1840. The sourcing of sap from the Malaysian *Palaquium gutta* tree marked the start of the engineering journey that would result in an economically sustainable subsea cable industry.

Gutta percha gum[43]

Gutta percha gum was first imported into Britain in 1843. Initially used to make decorative items such as chessmen, it proved to be an excellent insulator, initially for terrestrial telegraph systems (including underground cable) and then for subsea cable. Usefully for subsea cables, its insulating properties improved under pressure and at low subsea temperatures; gutta percha absorbs water, but in salt water the absorption rate is slower. Cables were therefore stored in a solution of brine from manufacture through to the laying process. By a process of trial and error and observation, the best option was to keep the brine at a steady 50°F. The Gutta Percha Company was set up in 1845 by John and Jacob Brett; (who laid the Dover to Calais cable in 1851) and dominated cable manufacture and deployment through the 1850s. By 1861 they were importing 1,000 tons of gutta percha per year, and by 1865 the company had supplied 14,000 miles of cable core for 64 cables. Gutta percha was used into the 1930s, when it started to be replaced by polythene.

The first transatlantic cable in 1856 had seven conducting wires wrapped in three coats of gutta percha wrapped, in turn, in tarred hemp (Wheatstone's idea) then an 18-strand sheath of iron wires. By 1866 these cables were transmitting across the Atlantic at six to eight words per minute at $5 per word, at 1866 dollar value; even at a deployed cost of $1 million, the cables provided a good return. The *Great Eastern* was decommissioned in 1888, having laid a total of 30,000 miles of cable. By 1900 there were ten Atlantic cables, with ten more under construction; there were 36 ships laying new lines and repairing 130,000 miles of legacy cable. By 1914 the total length of cables had increased to 300,000 miles.

The other Penders – the sons and grandsons of the Cable King

There is a whole room at the PK Museum dedicated to the history of John Pender (the Cable King) and his family, and their respective roles in building a subsea cable empire that connected the Empire via the telegraph hut at the top of Porthcurno Beach (more of this later).

John Cuthbert Denison-Pender (1882–1949), the 1st Baron Pender, and John Jocelyn Denison-Pender, the 2nd Baron Pender, the son and grandson respectively of John Pender, continued to be associated with the company through the first half of the 20th century, including the years immediately before and after the establishment of the Cable & Wireless Company in 1933.

In the early 1950s the government took control of the company, with all board directors appointed by the government. Speaking in 1950 at an event at the Science Museum to mark the centenary of the laying of the first Dover–Calais telegraph cable, the 2nd Baron Pender dispatched a message which was cabled around the world in 53.6 seconds. Given that in the 1930s the Marconi Company could send a shortwave beam radio message round the world in seven-tenths of a second, this was perhaps not as impressive as it might have been. That said, three generations of Pender had helped to steer the subsea cable industry through 100 years of technology and commercial innovation.

John Cuthbert Pender was appointed Tory MP for Newmarket in 1913 and MP for Tooting in 1918, carrying on the family tradition of representing cable interests in Parliament.

Wet assets and dry assets – subsea cables in Porthcurno

In today's subsea cable industry, a distinction is made between wet assets (the assets under the sea) and dry assets (the buildings on dry land which house the termination equipment, switches and data centres). In 1870 the arrival of the cable from Carcavelos in Portugal, completing the link from Bombay, meant that dry assets needed to be built in the Porthcurno valley to service the newly arrived wet asset. In modern terms, Porthcurno was in the process of becoming a communications portal connecting Cornwall (and London) to the rest of the world. The data (telegrams) had to be moved along cables up from the beach, to be read/decoded, then gummed onto tape (the state-of-the-art recording and storage system in the Victorian cable world) then sent to its destination over the terrestrial telegraph network.

The original telegraph station buildings were situated at and around what was to become known as Zodiac House, with additional staff cottages added in 1873, when another cable was laid from Vigo in Spain. A third cable arrived at Porthcurno from the Isles of Scilly in 1878 and a fourth from Gibraltar in 1887. The tennis courts were established on rented ground in the 1880s; telegraph operators and engineering staff needed some form of relaxation other than drinking. Mercury House, providing separate accommodation for the superintendent, was built in 1896.

In 1880 the Compagnie Française du Télégraph de Paris à New York (renamed the PQ Company after one of its directors, Monsieur Pouyer-Quertier), brought ashore a cable from Brest which formed part of the transatlantic telegraph route from Porthcurno to Nova Scotia. As the PQ Company was a competitor to the Eastern Telegraph Company, the cable was routed up the cliff to a hut (demolished in the 1950s) then linked by landline to a telegraph office in Penzance, at 55 Chapel Street. In 1918 a second Brest cable was laid. Then in the 1930s the Eastern Telegraph Company (ETC) took over the termination equipment and landline links to London. A white stone at the top of the cliff between Porthcurno Beach and Logan Rock marks the old site of the hut; some rusty iron wires in the rock mark the point on the cliff edge where the cable headed down to the sea as part of what was known at the time as Project Pluto, and a large yellow sign announces the presence of a telegraph cable.

With new cables arriving onshore prior to 1904, the telegraph station moved across the valley to new purpose-built fire-proofed accommodation at what was to become known as Eastern House.[44] In 1906 the arrival of another cable, from Faial in the Azores, led to Eastern House being extended northward.

Four years earlier, on the headland known as Pedu-Meor, which later came to be known as Wireless Point, a small two-roomed hut had appeared. Built by N. Holman and Sons of Penzance, the hut was connected to a 170-foot pitch pine mast towering above the cricket field. The hut housed 'undefined radio equipment' installed by Neville Maskelyne on behalf of ETC. Maskelyne was a professional entertainer and prestidigitator (magician) and dabbler in all things radio, with a brief to keep an eye on what Marconi was doing across the bay.

As we shall see in Chapter 6, Marconi's apparently successful attempt to send the letter S across the Atlantic by radio in 1901 had dented cable stock value, and although the board made light of the perceived threat, there was a measure of paranoia about the 'goings on' in Poldhu. The mast and the receiving equipment in the hut were destroyed by lightning in 1909, but radio monitoring by then had become, and remained, a useful discipline at Porthcurno.

Power-economic long-range radio was still a future threat rather than an immediate one. It was in the 1920s that Marconi's progress with shortwave beam radio marked the moment when the threat became real. The result was a 1928 conference held in London, recommending an amalgamation of all the cable and wireless interests within the Empire, with the long-distance radio assets of the Marconi Company. This led to the creation of a new operating company with the working title of the Imperial and International Communications Ltd.[45]

The following year, 1929, Porthcurno was accepted as the most important telegraph station in the British Empire, with 14 operational submarine cables terminating in the new concrete-built cable hut situated close to the beach. This was 'peak Porthcurno'; a significant percentage of the country's telegraph traffic was pouring into the sea 200 yards down the hill. The rubble and concrete building (free to visit; it is on your left as you go down the last bit of the valley from the car park) is described in the museum's guidebook as a

> small, single roomed, simple box shape, flat roofed structure with a wooden casement window in the East wall and central North facing door. Internally, a metal frame attached to the wall on three sides supports a series of 14 cable terminal boxes, each labelled with its destination and a commutation frame.

'Commutation' in this context is the process of reversing the direction of current flow in a circuit by switching the connections of the device, in this case the cable, to an external power supply.

The cables enter the building through holes in the floor adjacent to the wall and are joined to landlines within the cable terminal boxes. The landlines leave the box at the top and are carried underground to the telegraph station at Eastern House.

Cornwall connects the Commonwealth (via the hut on the beach)

What had actually arrived at the beach from 1870 onwards? The first cable to land at Porthcurno had been the one from Carcavelos in Portugal, in 1870. The second Carcavelos–Porthcurno cable was laid in 1873, with an intermediate landing at Vigo, Spain. The 1884 Lizard–Bilbao cable acquired by the ETC was first diverted into Falmouth and then Porthcurno. 1887 saw a further Porthcurno–Carcavelos–Gibraltar cable, laid by the CS *Scotia*, which in 1897 laid another cable, this time direct from Porthcurno to Gibraltar.

In 1901 CS *Anglia* laid a cable from Porthcurno to Madeira, to complete the link from Cape Town, and in 1906 CS *Colonia* laid a cable between Faial, in the Azores, and Porthcurno. In 1919 CS *Stephan* laid a direct Porthcurno–Gibraltar 'fast' cable, designed to be the first stage of a new 7,000-mile Far East submarine cable through the Mediterranean via Bombay and Madras to Singapore.

Laying cables had become a tried and trusted routine. CS *Stephan* had sailed from Greenwich loaded with just over 1,400 nautical miles of cable manufactured by Siemens Brothers. The first stop was Porthcurno, to drop the cable onto the beach. The engineer's report for the Porthcurno–Gibraltar No. 4 Cable states that there was 'a gentle north-easterly breeze' and that 'the cable end was floated ashore, supported by ninety wooden casks'. Two weeks later, the ship (and the other end of the cable) arrived in Gibraltar on a 'fine, bright and clear day. We completed final tests and advised the Managing Director'.

In 1925, the CS *Colonia* laid another Porthcurno cable to Bilbao. These cables were additional to a cable that had been laid in 1872 by the Direct Spanish Telegraph Company (owned privately by John Pender) which linked Bilbao to a landing point at Kennack Cove just below Goonhilly Down (where BT cables now connect Goonhilly to the subsea cable world). The 15-mile landline to Falmouth crossed several fields and was regularly ploughed up by farmers, incurring constant repair costs, so in 1921 the cable was diverted to Porthcurno. The cable earned revenue from cattle trading between Cornwall and Spain, adding cows to cauliflowers and cabbages as an income for the Telecoms Coast.

A few years later in 1934, the company changed its name to Cable & Wireless, which it retained until the beginning of the 21st century. The Cable & Wireless story is covered in Chapter 5.

Back to the future

In the meantime, we have been busy programming our time machine and are off to have lunch in the telegraph office canteen in 1895. The 80 or so operators have been busy all morning hand translating 2,000 messages from the three working cables and gumming the messages onto tape, and Simplex the shire horse is having another nosebag of oats after making the first of his two weekly trips to Penzance. Back into the time machine, this time to 1912, for a chat with pre-war holidaymakers watching the operators working those Edwardian-era cables, and then a 30-mile trip to Poldhu to meet Ambrose Fleming and Richard Vyvyan for an update on their latest research on high-power radio systems.

By 1911 there were 150 operators at Porthcurno and hundreds of operators working globally. The social and economic impact of the ETC for the few people living in or near Porthcurno valley and (as we shall see in Chapter 6), the social economic impact of the Marconi radio station in Poldhu, would have been significant for people living locally but relatively insignificant for Cornwall as a whole; in 1877 and in 1912, mining was still the dominant source of employment and economic activity in the county. However, Porthcurno and Poldhu were to play a pivotal role in connecting Britain with the Commonwealth (an institution which somehow still manages to survive today) and the rest of the world. As we shall see over the next few chapters, this legacy lives on along Cornwall's Telecoms Coast. My task is to highlight the pivotal moments in this process from the 1870s through to today.

Porthcurno, Poldhu and the Post Office

Our starting point is the Porthcurno and Poldhu postman, who still turns up once a day on most days (along with a motley connection of white vans with parcels ordered from Amazon).

We cannot move parcels through a terrestrial or subsea cable or across a radio communications system, but we can send packets of information and send and exchange written words and facilitate financial transactions. Within this context, the introduction of the Penny Post, largely at the instigation of Rowland Hill in 1840, established the underlying economic context of the telecommunications industry, including subsea cable. The business of the Post Office, after all is said and done, is the business of communication at a distance. It is a telecommunication system (the word 'telegraph' means 'distant writing' in Greek). In 1914 in London, the postman could arrive on your doorstep potentially 12 times a day, and it was a relatively low-cost service (the Penny Post was retained until immediately after the First World War). During the war, the postal system and telegrams were the two ways in which the few people left in Porthcurno and Poldhu would have kept in touch with the outside world.

After the First World War, the exponential growth of the telephone network began to impact on the revenue of the inland telegram business (still owned by the Post Office). At that time, unsurprisingly, there was a perception that a telegram delivered to your doorstep by a postman was going to be bad news – but research by the Post Office in 1934 showed that in a time of relative peace this was a misconception, and that less than 1 per cent of telegrams ended in tears, 4 per cent carried congratulations, 66 per cent were about the price of fish or food or commercial transactions, and 6.5 per cent were bets placed mainly on dogs and horses.

Mrs James gives the telegram a new lease of life

This led to the brilliant innovation by Mrs M.C. James of the 'greetings telegram', delivered in a gold-coloured envelope (a red envelope denoted something more serious). The Post Office would also keep a diary for you, reminding you about birthdays and other anniversaries for the price of threepence over the standard price of sixpence. With the inclusion of Christmas and Valentine's Day, the number of telegrams increased from 30 million in 1935 to 50 million in 1939, 4 million of which were greetings cards. We are straying at this point into the next chapter, but by the end of the Second World War in 1945, inland telegrams had increased to 63 million per year, with Cable & Wireless allowing parents and children to send one message a month for free via the 'imperial telecommunication system'. But in 1948 the Post Office Accounts were reporting a sixpence loss on each inland telegram delivered. This meant that prices had to increase, which in turn meant that traffic declined (to 34 million messages in 1954). The basic cost was then increased to three shillings. By 1964 the price had increased to five shillings, and volume was down to 10 million.

Losses had been partially offset by the introduction of the deluxe wedding telegram in 1957, the deluxe birthday and baby greetings in 1959, and the deluxe 21st birthday telegram in 1961. These, however, only delayed the inevitable; the last telegram in the USA was sent in 2006, the last UK telegram was sent in 2008 and the world's last telegram was sent from India on 14 July 2013.

The heavy non-interrupt flywheel in the Porthcurno telegraph office had been allowed to spin to a halt in 1970 and the last live subsea telegram cable was allowed to go forever silent. Fifty years on, Japanese, Korean and Chinese holidaymakers sip cappuccinos at the beach cafés in Porthcurno and Poldhu and upload high-quality video to their Instagram accounts via the local Vodafone cell site.

This miracle of modern communication is based on 150 years of technology and engineering invention. Porthcurno and Poldhu played a pivotal part in this innovation process. To understand how this all became possible, it is time to nip back into our time machine and head back to the 19th century via Ancient Greece.

The birth of the battery

School textbooks credit the Greeks for discovering electricity 3,000 years ago, and they are probably right. The Greek word 'elektron' means 'amber', the yellow fossilised substance that is produced from tree sap. Amber, when rubbed against wool, attracts straw or feathers; the modern-day equivalent is getting a balloon to stick to your jumper at a children's tea party or shuffling along a hotel carpet in your socks and getting a shock from the bedroom doorknob. Benjamin Franklin did that dangerous stuff with a kite in 1748 and described an array of charged glass plates as a 'battery' because of the way the plates were lined up in rows like an artillery unit. The glass plates were a Leyden jar, accidentally discovered by the Dutch physicist Pieter van Musschenbroek at the University of Leiden two years earlier (in 1746) and possibly the year before by the German inventor Ewald Georg von Kleist. The French were also involved; Abbé Nollet ran a monastery in Paris, with a side hustle as an experimenter with static electricity, connecting 700 Carthusian monks into a 5,400-foot circle of iron wire, a follow-up to a similar experiment with 180 royal guardsmen. Monsieur Nollet became a Member of the Royal Society of London and the first Professor of Experimental Physics at the University of Paris. Those monks and guardsmen were a useful career move.

In 1901 Ambrose Fleming's 'thunder factory' in Poldhu[46] used a two-stage spark transmission process – essentially a static electricity discharge system based on revolving discs with copper tips, producing anything between a few tens to a few hundred sparks a second; but the spark transmitters needed a steady source of electricity, as did the cables snaking out under the sea at Porthcurno.

One hundred years earlier in 1800, Alessandro Volta, Professor of Natural Philosophy at the University of Padua, had discovered that alternating discs of zinc and copper separated by pieces of cardboard soaked in brine would produce a steady electrical current. This was the world's first wet cell battery, and probably the first time that the conductive properties of salt water had been observed and at least partially understood. One volt was later defined as the potential difference between two points on a conducting wire corresponding to an electric current of one amp (named after André-Marie Ampère) dissipating one watt (named after James Watt) between those points.

In 1821 Michael Faraday (who had the farad, a measure of electrical capacitance, named after him) worked out that a magnet spinning within a coil of copper induced an electrical current that would flow through a wire and transfer an electrical charge. In one fell swoop he had discovered how to generate electricity efficiently and how to transmit it efficiently as well. Add a battery (to store the electricity) and you had the basic technology ingredients for long-distance terrestrial and subsea communication.

Most of the early telegraph systems were powered either by Daniell cells (invented by John Frederic Daniell in 1836) or Leclanche cells (invented by Georges Leclanche in 1866). The Daniell cell had a copper electrode immersed in a copper sulphate solution and a zinc electrode immersed in a zinc sulphate solution. The two half-cells were connected by a 'salt bridge', a porous material soaked in a solution of potassium nitrate. The Daniell cell produced a relatively stable voltage of about 1.1 volts, but was bulky, and the solutions needed to be continually topped up; in addition, unwanted by-products built up and stopped the cell working.

The Leclanche cell used a zinc electrode in a zinc chloride solution and a carbon electrode suspended in a mixture of manganese dioxide and carbon powder which acted as a depolariser. The two electrodes were separated by a porous pot that acted as a separator. The Leclanche cell was simpler than a Daniell cell and easier to maintain but was not rechargeable.

In 1859 the French inventor Gaston Plante developed the first practical rechargeable lead acid battery. In 1867 Wheatstone put the finishing touches to a working dynamo based on Faraday's earlier work, so batteries could now be recharged. Siemens had a dynamo working by then as well.

Thomas Edison invented the alkaline battery in 1901, then – we need to move on, as this is not a history of batteries – the lithium battery arrived in the early 70s. After 50 years of intense engineering the lithium-ion battery has delivered the smartphone, which lasts miraculously long on a single charge, and a new generation of (relatively) environmentally friendly electric cars. Lithium is being promoted as the next mining revolution in Cornwall.[47]

Batteries and link budgets

Engineers designing radio and wireline systems in the 20th and 21st centuries do a link budget to work out how much power is needed to transmit at a certain speed over a certain distance based on a calculated noise floor and considering the signal distortion introduced from one end of the link to the other. In the early 1870s, however, it was more a case of trial and error.[48] The cable laid in 1878 from the Isles of Scilly into Porthcurno provides an example. The cable replaced a short-lived cable that had been laid directly into Land's End in 1869. The notes in the commissioning log for the day record that once the cable had come on shore it was connected in less than an hour and that 'the people of Scilly were very pleased to be put through to Penzance – we are only using 20 cells' (suggesting that 30 volts was needed to close the link).

Going underwater

In an ideal world, the subsea cable industry would have taken the technologies used in the terrestrial telegraph industry and just put them under the sea. To an extent this was possible, as both systems relied on the deflection of a needle by a voltaic current to read the messages encoded on to the cable. Various systems were developed in the 1820s, but it was Cooke and Wheatstone on one side of the Atlantic, and Morse and Vaile on the other, that produced and sold and installed the first practical systems (back to Brunel and his railway at the end of Chapter 3).

Cooke had a working prototype of a needle telegraph that in November 1836 he showed to Michael Faraday, who more or less approved, stating: 'the principle of the instrument was right'. But Cooke felt the need for a partner with broader scientific and manufacturing experience, hence the 1837 proposal to Wheatstone that they take out a joint patent. Like many of the 'men of science' of the 18th and 19th centuries, Wheatstone was involved in many different things, including the design and manufacture of musical instruments. He was an expert on the acoustic principles of violins, flutes and trombones, and is credited with inventing the modern accordion. Cooke and Wheatstone's agreement with the GWR, finalised in April 1838 and confirmed on 24th of May, marked the start of the rollout of telegraph lines along railway tracks, initially with cables laid within an iron tube next to the track then via poles supporting suspended wires, with glass and pottery insulators.

On 24th of November the General Traffic Committee of the GWR agreed that Cooke could transmit messages for the general public in addition to supporting the signalling needs of the railway. The signalling was a significant success; it meant that trains didn't run into each other and could safely share a single track when required via a system developed by Cooke known as the 'block system'. Railway safety provided a useful boost to the telegraph industry. Fifty years on, Safety of Life at Sea was to be similarly useful to Marconi.

However, while things were going more or less according to plan on the railroads, it was definitely not plain sailing for the subsea cable pioneers.

Cables under the sea – oceans and oceanography

Early cable failures were partly due to badly designed and poorly manufactured cable, but also due to the immense depths; the cables had to be capable of withstanding both extreme pressure and temperature, as well as the mechanical strain when cables transitioned from the continental shelf (typically at a depth of 200 feet or so) to the deep ocean. In addition, the deep ocean floor is far from flat. The first cable lost in the Atlantic Sea was at a depth of 2 nautical miles, while parts of the Mariana Trench are at a depth of well over 6 nautical miles; the hills are higher, and the valleys lower under the sea than they are on the land masses in between.

The success of the first international submarine cable between Dover and Calais, which went live on 13 November 1851, instilled an early confidence based on the fact that the cable worked and made money; for example, the price of securities on the Paris Bourse would be known on the floor of the London Stock Exchange on the same day and within business hours – but the 21-mile route under the Channel was shallow all the way. The 1851 Great Exhibition had 13 telegraph instruments on display, indicating a wave of enthusiasm for this new modern miracle.

The ETC, with the ubiquitous Pender as a board director, won a concession from the Dutch government to lay cable from Orford Ness on the east coast of England to Scheveningen in the Netherlands, and on to The Hague, with two cables laid in 1852 and 1853. A new company, the International Telegraph Company, was formed, partly to smooth relations with the Dutch government and Dutch investors (one third of the shares were held in Amsterdam, Rotterdam and The Hague), and by the mid-1850s London was in direct cable contact with all the major continental cities.

Connecting to the Commonwealth was a bigger challenge. The cables from Porthcurno heading down the French and Portuguese coasts and onwards either to the East though the Mediterranean or down the east coast of Africa, or across to Latin America and Brazil, had to be laid on a sea floor every bit as hazardous as the route from the west coast of Ireland to Newfoundland. An overland cable route had been established to India in 1865, but a sea route with safe and secure stations on the way was regarded as strategically essential (a political imperative) and, with end-to-end control of the cable, potentially more profitable and secure.

Commercial investment, technical innovation, commercial innovation

Commercial investment spurred technical innovation. Wheatstone's ABC Telegraph, patented in 1858, helped to make the life of the operator easier, and increased the number of words per minute. The invention earned Wheatstone the princely sum of £9,200 from H.M. Government as part of the privatisation process. By 1868 the public telegraph network of the UK consisted of over 91,000 miles of wire and 21,751 miles of line operated by railway and telegraph companies, with 3,381 stations open to the public transmitting 6 million inland messages, while the telegraph companies owned the greater part of the wires and lines, and provided the bulk of facilities. Public facilities were provided by railway companies at 1,226 of the stations. The railway stations were co-sited with approximately 36 per cent of all telegraph stations, but users found it generally more convenient to pop into their local post office. That said, at the end of the 1860s the railways were hosting 55,000 miles of telegraph company wire, which would provide an income for years to come (the routeing of fibre cables still provides useful income today). It was these assets that were nationalised, and the timing could not have been better for Pender; the cash from the buyout of his terrestrial telegraph shares arrived just in time to fuel 30 years of growth in his long-distance subsea cable business, with at least some of that investment centred on Porthcurno.

Terrestrial versus subsea

While some things were easier on the long-distance subsea routes, some things were harder. Once the government had taken over control of the terrestrial cable network it could, theoretically at least, enforce access for cable routes through cities and towns and across the countryside – but, like power pylons today, telegraph poles were never popular with the public. Worse, the poles did not last long; native larch or pine imported from Sweden was coated in tar, but they only lasted for seven years (today's pressure-treated telegraph poles, by comparison, last the best part of 50 years); and sulphur from city smokestacks converted the oxide in the iron into a sulphate of zinc, which dissolved in the rain. As a result, maintenance costs could easily account for 30 per cent of the total working cost of the network.

The perils of the deep blue sea floor

Going under the sea was (and still is) simpler in some ways. It doesn't rain under the sea, though it is wet. Once you are in international waters you don't need to worry about who owns the next nautical mile of ocean floor. That said, the cable between Porthcurno and Lisbon is not untypical of a route that at times is nearly 2,700 fathoms deep (2.7 nautical miles),[49] one of the greatest depths in which any cable had been laid in the 1870s and subsequently the deepest depth at which a cable had been repaired.

The Atlantic route from the west coast of Ireland had been planned on the basis of a series of soundings taken by Lieutenant O.H. Berryman of the US Navy in the USS *Arctic* and by Commander Joseph Dayman of the Royal Navy in HMS *Cyclops*. They discovered (or thought they had discovered) 'a gently undulating plateau of great breadth, at a depth varying from 1,700 to 2,400 fathoms (1.7–2.4 nautical miles) extending nearly the whole distance between Ireland and British North America, comparing favourably with soundings of 6,000–7,000 fathoms (6–7 nautical miles) further south', hence the choice of Trinity Bay in Newfoundland and Valentia Bay in Ireland as the start and end points. (See Appendix 4; the plateau was largely mythical).[50]

In 1876 the invention of the bathometer by William Siemens enabled the depth of the ocean floor to be measured without the use of a line. The instrument, based around a tube of mercury, was similar in principle to the barometer; the mercury, acting under the force of gravity, pushed down on a small steel sheet. The height (and therefore pressure) of the mercury was proportional to the strength of the Earth's gravitational field, the theory being that on the basis that water has a lower density than the earth's crust, the greater the depth of water under the ship the lower the gravitational force. Once calibrated, these instruments worked adequately well.

Making cables last longer under the sea

In 1859 the British government had decided to appoint a committee to consider the question of the construction, laying and maintenance of submarine cables. The committee included Charles Wheatstone, Sir Joseph Thomson and Joseph Whitworth, the man who standardised the screw thread.

There were processes that needed to be improved. Terrestrial cables could rely on air as an effective insulator. By comparison, thanks to Volta, there was an understanding that sea water is an effective conductor with the cable core acting as a capacitor, a sort of underwater Leyden jar. Gutta percha was an excellent insulator (and improved with increased pressure and cold), but it needed to be incorporated into the production process with care and attention, with the finished cable stored in brine then coiled into tanks in purpose-built cable ships (or carefully adapted ships such as the *Great Eastern*) then expertly laid. Even then, all sorts of things would go wrong, requiring fleets of specially equipped repair ships.

This was one of the secrets of John Pender's success. Like Elon Musk with his rocket and satellite business, Pender set out to control all aspects of the cable manufacture, cable laying and cable repair business through his separate Telcon Company. The production plant in North Woolwich covered over 16 acres, and produced cable, insulators, electric and magnetic apparatus, with galvanising and rolling mills, and steel and wire works. The works in Silvertown (now a station on the Docklands Light Railway) covered 15 acres and employed 2,800 people, and was one of the first factories to be lit by electric lamps.

The trials and tribulations of maintaining a subsea cable system are recorded in detail in the Board Meeting Minutes and internal technical studies stored in the archive at Porthcurno. John Moyle, the author of *Cornwall's Communications* (see Acknowledgements) spent many hours reading and analysing these documents and concluded that maintenance costs were higher than the company either wanted or was prepared to admit to, partly due to concern for investor confidence and the potential impact on the stock price of the company.

Cables going slowly under the sea

The cables were also comparatively slow. Transmission speeds on the transatlantic cable route (1857–1858) were three to seven words per minute (wpm), an order of magnitude slower than landline speeds. But by the time of the First World War speeds of up to 80 wpm were typical. Faster speeds could have been achieved by using thicker copper cable, but it was generally more economic to improve receiver equipment. Bigger cables cost more to make but were also harder to coil and store in a ship, and harder to handle. Bear in mind that subsea cables were more expensive to manufacture than the bare cable used in terrestrial telegraph systems.

In America, the use of Morse's eponymous dot-dash had become widely used on terrestrial telegraph systems (though other coding systems were used as well) but Morse's system did not work well under the sea, as the dots became blurred with the dashes, which effectively limited the speed of transmission.

In the 1850s two French scientists, Fizeau and Gounelle, measured the speed at which electricity travelled through copper and iron, coming up with an estimated figure of 179 million metres per second in copper, and 100 million metres per second in iron wire. (To put this into context, radio waves (electrons) and optical waves (photons) in free space travel at the speed of light (300 million metres per second); in an optical fibre cable the photons fly along at a relatively pedestrian 270 million metres per second.) By comparison, Victorian subsea cable was even slower. Faraday and Clark had estimated the transmission time through 1,500 miles of subterranean wire to be two seconds, the reduced speed due to the inductive capacitance caused by the sea water (or subsea saline sludge when cables were buried under the sea floor). The effect was different, depending on sea depth and density and saline concentration. Improving the insulation helped but generally made the cable bigger, heavier and harder to handle.

The workaround was to use a modified version of Morse code: Cable Code. Instead of a single electrical polarity (as used in terrestrial telegraphs), signals in subsea cables used both polarities (positive and negative). At the transmitting station, a pulse of electric current was keyed into the insulated single-core copper cable: positive for a dot and negative for a dash. A special form of Morse key was used, one half generating dots and the other generating dashes, the operator using the index and middle fingers respectively of one hand. At the receiving end of a submarine cable, the operator used a galvanometer to observe the fluctuations in electrical current which formed the signal code. The circuit was completed by using the sea water for what was known as the 'earth return'. From 1894 increased diameter cables were used. It was these cables that were described as 'high speed lines'.

Speed, accuracy and cost

There were other ways to improve throughput and efficiency, including duplexing (the ability to send messages up and down a cable at the same time), introduced from the late 1870s onwards, and various mechanisms devised to automate the encoding and decoding process. These were introduced out of necessity after the First World War due to so many men dying in the war and from the influenza epidemic. By 1918 cable traffic was being encoded and decoded in London, not Porthcurno. Further down the line, synchronised electronic mechanical relay regenerators (REGEN) started to be installed from 1925, reducing costs and improving accuracy; taking operators away meant that operator mistakes, the equivalent of Chinese whispers, were no longer an issue.[51]

Generally speaking, the number of repair ships needed increased at the same rate as the expansion of the cable network, suggesting that the overall reliability of the system only improved slowly, so other cost-reduction opportunities were comparatively important.

That said, from the 1850s onwards there were steady improvements in the techniques used to determine where cable faults were happening. From the late 1850s Charles Tilston Bright started to use calibrated resistance measurements, with cables tested at the end of the production line and then retested through the laying and commissioning process. The cable between Malta and Alexandra in 1861 was an early test of the process; the location of a cable fault could then be estimated by the change in resistance compared to a known reference.

In the 1860s the standardisation of the ohm as a unit of resistance was a big leap forward. William Preece, then a young and rising engineer at the Post Office, helped to improve the accuracy of fault detection, pointing out that although electrical resistance was proportional to the length of the copper core of the cable, the resistance was determined by the uniformity of the core (including the purity of the copper). The good thing about this was that as the production processes improved so did detection accuracy, which meant that cable repair ships could be sent to where they were needed with reasonable precision.

Improved cable quality therefore had two benefits: more accurate fault location, and faster throughput. Technical papers in the PK archive describe this as occurring in three stages: low-speed lines from 1873 to 1882, high-speed lines from 1894 to 1923, and very high-speed lines from 1923 to 1940.

From the Second World War onwards, the scope and appetite for further optimisation of subsea telegram cables reduced; 100 years of engineering optimisation had increased transmission speeds from a few words per hour in the late 1850s (the Ireland to Newfoundland route using Morse code and the Thomson galvanometer being the reference point), to six to eight words per minute by the late 1860s (Morse plus the Siphon recorder, a pen and ink device, on the Ireland to Newfoundland cable), to 40 words per minute by the late 1890s (cable code with automatic transmitting equipment) and 100 words per minute (2,500 characters) by the late 1920s (cable code on the Newfoundland to Azores cable). In the 1950s the first transatlantic subsea coaxial telephone cables and powered subsea repeaters marked the beginning of the end for the subsea telegram business – but that is a story for the next chapter.

Testing times in Porthcurno, and a training opportunity

By the turn of the century an extensive literature was available for test engineers. This body of work established the basis for engineering training, which became an important part of the Porthcurno story through to the early 1990s. From 1870 onwards trainees had been sent down from London after 12 months of initial training to spend another six to twelve months getting to know how to operate and maintain a subsea cable system.

The cable station was used to train maintenance engineers looking after the batteries (see Appendix 3) and transmitting and receiving equipment. In the 1960s a line of telegraph poles was built between the college and the beach so that engineers could learn how to install, test and repair terrestrial telephone lines. The facility is still there on your left as you walk down the valley. Before cars came to the valley, if the station went offline for any reason messages would have to be sent on foot or horseback to and from Penzance. When the Great Blizzard of 1891 brought down the telephone lines, two volunteers had to walk from Porthcurno with the day's messages and nearly perished in the snow.

Engineering training was separate from operator training. Trainee operators were known as probationers and received no salary until they passed their exams. By 1900 the training had become more formalised; being posted overseas as an engineer meant that you had to know enough about more or less everything to be able to fix more or less anything. In 1888 the founding of the Camborne School of Mines marked the beginning of a similar process in the mining industry, training engineers who would go to work in mines in every corner of the globe.

Although Porthcurno no longer runs telecommunications engineering training courses, there is still a worldwide demand for telecommunications engineers, and the same can be said for mining.

Even with the railway to Penzance, Porthcurno was a long way from London, so a posting there was likened to being sent into exile. In the 1920s a sports and recreation centre was set up in Twickenham in West London called the Exiles Club; it remained open until the 1990s.

John Pender says goodbye

I need to decide when to finish this chapter, and the death of John Pender in July 1896 seems to be an appropriate, though poignant, end point. Pender had created a cable empire that at its peak would employ 54,000 people, with cable stations (and radio stations and satellite earth stations) all around the world.

In 1870, when the cable from India went live in Porthcurno, the time taken to get a message from London to India reduced from 6 weeks to 11 minutes. Hugh Barty-King illustrates the Pender principle of profit, philanthropy and paternalism in his 1979 book on Cable & Wireless, *Girdle Round the Earth* (a quote from Puck in Shakespeare's *Midsummer Night's Dream*),[52] referencing John Milton's 1667 *Paradise Lost*: 'Good, the more communicated, more abundant grows'.

A good dose of politics needs to be added into this mix as well, but the two quotations set the stage for us as we transition from the 19th to the 20th century, to plot the rise and fall of the Cable & Wireless Company.

5
CABLE & WIRELESS

Cable or wireless

I could have started this chapter in 1934, the year in which the Cable & Wireless Company was established, or in 1929 when its predecessor, the Cables and Wireless Company, was set up alongside the Imperial and International Communications Company. Instead, I am starting in 1896 with the death of John Pender, aged 80, and the appointment of John Denison-Pender as Managing Director of the Eastern Telegraph Company (ETC).

Over the next 30 years the 30 telegraph companies that John Pender had established would be consolidated into the single entity that would become Cable & Wireless, with Guglielmo Marconi as a somewhat reluctant and resentful company director.

That journey had also started in 1896, with Marconi's arrival in London from the Marconi country estate near Bologna in Italy, accompanied by his formidable mother, Annie Jameson, scion of the Jameson Irish whiskey family.

Like the Brunels 60 years earlier, Marconi was a master of the art of public relations, and by 1900 his adventures in the exciting new world of radio were reported in the national and international press on an almost daily basis. It would be 20 years before long-distance radio became a serious competitor to long-distance subsea cable, but that didn't stop the hedge fund managers of the day talking down the stock of the ETC.

To an extent the ETC board members were reassured by the technical and commercial challenges that Marconi faced in those early years; the practical shortcomings of radio were highlighted in the Boer War, with equipment shipped but not used.

The Boer War: subsea to South Africa[53]

For the ETC, the Boer War went rather well. Mining might be going through difficult times in Cornwall, but down in the Transvaal Cecil Rhodes was making spectacular sums of money from digging up diamonds and mining gold. The mines needed railways, the railways needed telegraph networks, and the telegraph networks needed to be connected to London.

Cecil Rhodes had been born in Hertfordshire in 1853, and he died on 26 March 1902. In 1870, to improve his health, he was sent to join his older brother Herbert on his cotton farm in Durban with £3,000 that his aunt had lent him. The diamond fields of Kimberley had been discovered in 1866 and both Herbert and Cecil made the trek 650 kilometres north from Durban to see if they could make money from mining. After returning briefly to Oriel College in Oxford, Cecil went back to South Africa and founded the De Beers Company. By 1888 he had established a monopoly over the Kimberley diamond fields and controlled 90 per cent of the world's diamond production. His interests in the Transvaal gold fields were bringing him an additional £400,000 per year.

The articles of association of the company supported an ambition to extend the mining business northwards to exploit what was thought to be a gold reef running up the spine of Africa. Rhodes had a plan to build a railway and telegraph network to connect the Cape to Cairo. But the gold reef proved to be a figment of imagination and the construction of a 4,500-mile railway and telegraph network overland through the heart of Africa proved to be an impossible dream, which was good news for Porthcurno and the ETC.

Cecil Rhodes returned to South Africa from a trip to England in 1890 to become Prime Minister of Cape Colony. He backed the disastrous Jameson raid in 1895, when a small force of British soldiers attempted to overthrow Paul Kruger, the Afrikaner president of the gold-rich Transvaal Republic. This was the 19th-century equivalent of sending a British army to Ukraine to get access to Ukraine's rare earth mineral reserves, and it marked the start of a war in which tens of thousands of people would die.

In 1890 ETC had two cable routes from Porthcurno down to the Cape: either a fragmented route down the west coast of Africa, with vulnerable repeater/relay stations, or a longer route via Gibraltar, Malta, Alexandria, Suez, Aden and on down the east coast of Africa.

With Rhodes' overland telegraph route foundering, ETC despatched CS *Anglia* to Cape Town with 2,000 miles of cable. CS *Seine* was sent with another 800 miles of cable to St Helena, then on to Ascension Island and across the Atlantic to St Vincent, where the cable was connected to the existing St Vincent–Porthcurno link. The end-to-end cable went live on the 21 February 1900. This was arguably the first time that there was a widespread recognition that a reliable and secure subsea cable connection (preferably several connections) was an essential part of planning and supporting a military campaign 6,000 miles from London.

The First World War and the Cable War

On 31 July 1914, as tensions rose between Britain and Germany, a detachment of 43 soldiers arrived at PK and were billeted in the theatre. On the basis that Germans could soon be making their way into Mount's Bay, trenches were built on the beach, barbed wire was rolled out, windows were sandbagged and defensive blockhouses were built. Cable operation and maintenance was classed as a protected (defence-critical) profession. The government requisitioned the ETC offices in London, and suspended the use of message encryption for all commercial and private traffic. On 4 August, the day that Britain officially declared war on Germany, a censor started work at Porthcurno; in addition the Senior Inspector of Telegraphs arrived and arranged for the radio mast (which had been rebuilt after a lightning strike in 1909) to be taken down and for the radio equipment to be buried in hidden caches around the valley.

The Cable & Wireless War

Within hours of the outbreak of war, cable ships owned by the ETC, together with some Post Office ships, were despatched to cut the five cables of the German Atlantic Telegraph Company (DAT) and the German South American Telegraph Company at the point where they entered the English Channel. This meant that Germany could not cable to the USA or Britain. Additionally, the German landlines to the Far East were blocked, as they passed overland through Russia and India.

This was inconvenient for the Germans – but not fatally so, as the high-power longwave radio station at Nauen was by then fully operational, having been initially commissioned in 1906. There was a second high-power station in Hanover. The Australian navy captured the German wireless stations in Samoa and Nauru in the Marshall Islands, Herbertshohe on Neu Pommern island and in the Cameroons. The Germans retaliated by attacking Britain's Pacific and Indian Ocean cables routed through Fanning Island and Direction Island, Cocos. These cables were repaired a month later.

Marconi's longwave transatlantic transmitter at Cefn-Dhu, east of Caernarvon, known as the Waunfawr transmitter, had been commissioned five months before the outbreak of war, so, like the Germans, Britain had long-distance wireless links as an alternative to the long-distance subsea cable routes, though the transcontinental links, for example to Australia, would only go live after the war.

On 12 May 1914, at an annual general meeting of the Eastern Extension Company (the company established to manage cable traffic to Australasia and China), it was noted to applause that there had not been 'any loss from wireless telegraphy'. What ETC and its subsidiary companies failed to appreciate or anticipate was that the war would have a pull-through effect on radio component technology, particularly valve technology.

Radio equipment had made little or no difference in the Boer War, but it made a significant difference in the First World War for all the armed forces (the army, the navy and, towards the end of the war, the air force). This in turn improved the efficiency of longwave radio after the war, but more importantly facilitated shortwave radio, a technology that would transform the delivery economics of long-distance radio communication (the ability to send radio messages round the world). RF direction finding and radio signals intelligence (SIGINT) were also crucial to the war effort, but that is a topic for Chapter 8.

A cable boom and a problem

During the war, ETC sent any message regarding British killed and wounded for free, and sent telegrams between soldiers, sailors and nurses in the Expeditionary Forces and their relatives at quarter rate – but even with this additional cost, its profits increased by leaps and bounds; in 1915 the Eastern Extension paid an 8 per cent dividend free of income tax. Receipts were so high that £250,000 had to be put into a reserve fund. The Board even sanctioned that £30 18s 3d should be spent on wireless telegraphy, indicating that its members were still unconcerned about the threat of radio technology. The complacency was in a way understandable, because the cash just kept coming in. It was the same for ETC, which went from a turnover of £1.5 million in 1913 to £3.3 million in 1919.

A small amount of this money went into training at Porthcurno, which had been placed on a more formal basis. In 1915 an 'advanced' training course was established for 12 people every month, chosen from a class of 30 at the Electra House Training School in London, who headed down to sample the delights of the Cornish coast. Life in Porthcurno was still basic, with candles in the bedrooms, though electric light had been installed into the dining hall, sitting rooms, library and theatre, a modest investment considering that the Board in London were commissioning new silver and fine china for the directors' dining room.

As always, when a business seems to be going swimmingly well, there are problems brewing under the surface – or in this case under the sea. During the war the government had priority over all other traffic on the subsea cable networks. This was suspended in June 1919 – but no new cables had been laid during the war, and traffic had increased exponentially. Between 1913 and 1918, India Far East Traffic had increased by over 300 per cent, and at the end of the war a commercial telegram from Calcutta to London was taking 17 days.

So in 1919, a captured German cable ship, CS *Stephan*, one of the largest cable ships in the world, was sent to lay a 1,200-mile 'fast cable' from Porthcurno to Gibraltar, but this only partially solved the loading problem.

A partial answer

Part of the answer was to take operators out of the loop – and anyway this was unavoidable, because so many had died in the war or in the influenza epidemic. ETC had started to employ women telegraphists in June 1918, so this helped, but didn't solve, the labour shortage issue or the (related) problem of slow traffic throughput.

The impact on Porthcurno was immediate, with manual operation ceasing in March 1918. This was never reinstated. Porthcurno had become a relay station. This meant that not only were there no operators at the station, but there was no live equipment on which probation operators could be trained. Electra House reintroduced a short technical course in Porthcurno in 1919 on 'Recent Methods of Working', covering Cable Code and the change from sectional to through working, but this was short-lived, and in January 1920 a new training school was opened in Shepherd's Walk in Hammersmith.

Some new cables were laid into Porthcurno in addition to the Gibraltar cable, and old cables were diverted, including the PQ cable from Brest and a cable from Bilbao redirected from the Lizard. This was complemented by the new Porthcurno–Bilbao cable in 1925. This cable and gradually other cable routes were converted to Regenerator working from 1925 onwards, which helped to reduce end-to-end delays (the 17 days from India) but also reduced the need for operator training, operators and operator support staff.

A problem solved?

In the meantime, Marconi was making waves on the other side of Mount's Bay, and the wash from the waves was beginning to inflict damage on cable company share values. The Imperial Wireless and Cable Conference in 1928 was dressed up as a meeting of the great and the good to secure the best interests of the Commonwealth and Britain's future as an international telecommunications powerhouse, but there was skulduggery lurking beneath the surface, some of which was based on whether this 'best interest' could reliable be placed in the hands of a 'foreigner' – in this case an Italian-born man with a very English (or more specifically Irish) mother. Some dodgy share dealings in 1912 involving Lloyd George were dredged up and talked up in dark corners. This was malicious nonsense, but nonetheless effective.

Both ETC and Marconi had done well out of the First World War. Marconi had done well after the war mainly because he ran a remarkably adept company remarkably well and could make fast decisions. He was also charming and persuasive – and, to use a modern term, media savvy. He made the Pender Empire seem ponderous. ETC and its subsidiary companies had cash flows during the First World War which, viewed dispassionately, were akin to war profiteering, but this did not stop them from presenting themselves as the safe pair of hands that would bind the telecommunications industry to the future of the Commonwealth and the remnants of the British Empire.

By 1927, however, the cable industry had lost half of its international business to shortwave international relays. The outcome of the conference was the creation of a new company 'to merge all the international communications of the British Empire', which included all of Marconi's long-distance wireless businesses. Marconi stepped down as Chairman of 60 companies in 20 countries, a wireless empire of equivalent complexity and global reach to the cable empire that John Pender had created 30 years before.

Even as a minority shareholder, Marconi was a wealthy man with an assured income of £20,000 a year and £5,000 to spend on research, but the associated cost was a heart attack and failing health, which he would have to battle for the next ten years; he died in Rome on 20 July 1937. The Marconi story is told in Chapter 6.

The other outcome of the conference was the addition of two more cables into Porthcurno, purchased from the British Post Office, redirected from their original landing point in Mousehole. One of these was the cable equivalent of the Ancient Mariner, an 1874 cable manufactured by Siemens Brothers, originating in Harbour Grace in Newfoundland.

The other was a German cable from 1900 laid to Faial in the Azores. Porthcurno now had 14 operational cables. As J.E. Packer points out in his *Porthcurno Handbook*, staff could boast that they now worked at 'the most important cable station in the British Empire'.

Impact of the Second World War on Cable & Wireless and Porthcurno

In the 1930s the commercial impact of declining cable traffic was offset by investment income and the heroic efforts of Mrs M.C. James and her greeting cards (see Chapter 4), but essentially the telegraph cable industry was a dying business. On 1 September 1939 the Second World War was declared, and the cable industry reverted to being a dying business in an altogether more macabre way. In its place, including radio, Cable & Wireless traffic increased by 400 per cent over the next five years. In 1944 its profits doubled to nearly £2.4 million.

The company's shortwave radio station in Bodmin (Marconi's legacy, see Chapter 6) was taken over by the Air Ministry. The plan was that if the transmitters at Ongar and Dorchester and the receivers at Brentwood and Somerton were put out of action, Porthcurno would take over and operate whatever was left of the Cable & Wireless Network. If Cornwall was invaded, the Porthcurno and the US cables landing at Sennen Cove were to be disabled.

At the end of 1939 300 Scottish infantry arrived at Porthcurno and were quartered in the Old School. The beach was sealed off with wire and an unscalable floodlit fence, tank traps were dug, and emplacements were built for rocket projectors and a Bofors gun.

After becoming Prime Minister in May 1940, Winston Churchill instructed work to begin on an underground tunnel. On 25 June Edmund Nuttall Sons and Co., the Manchester-based company that had won the pre-war contract to build the British end of the Channel Tunnel (originally surveyed by Henry Marc Brunel, see Chapter 3) arrived with a skilled team of 200 Cornish miners from Truro, with some extra experts from Ireland and Yorkshire, and started tunnelling into the cliff rock, creating two tunnels 26 feet wide and 150 feet long, topped by a 70-foot granite roof reinforced with concrete; 15,000 tons of granite were removed by being loaded onto a narrow-gauge railway which had been built in the valley.

A 119-step emergency exit was built in case the blast proof doors jammed. Visitors today can make their way up the stairs to emerge into the constant Cornish sunshine at the top of the cliff above the valley. Into the cavernous space at the bottom of the stairs went the circuits, an instrument room and a generating room, most of which are still there (though not operational).

Cable & Wireless paid for the work to be done. This was a masterstroke of diplomacy. H.M. Government was entitled to take over possession of all the company's stations and cables on the outbreak of war but instead decided to leave the company in the capable hands of its Court of Directors, including the Lords Pender and Inverforth.

The tit-for-tat cable wars started again, and bombs falling on Hammersmith twice took out the lines to and from London, but Porthcurno remained comparatively peaceful, with just a few bombs falling harmlessly on local farmland.[54] [55]

Post Second World War: Porthcurno and the Telecoms Coast (the Commonwealth Coast)

Labour's landslide election in July 1945 marked the start of six years of nationalisation and the creation of the Welfare State. Technically the Cable & Wireless Company was government-controlled rather than privatised, but it was more or less unchanged.

On 1 April 1950, 3,500 Cable & Wireless staff joined the Post Office. The government now owned the 11 telegraph cables snaking from the beach in Porthcurno and stretching 155,000 nautical miles (186,000 statute miles) out to sea. The network included the world's longest cable, the Old Pacific Cable, stretching 3,460 miles from Vancouver to Fanning Island, plus 132 telegraph cables, 105 radio telegraph circuits, 66 radio telephone circuits and 10 radio photograph telegraphy circuits.[56]

Cable & Wireless Holdings retained office space in London, a cable depot in Plymouth, and Porthcurno, where a demarcation line was established half a mile inland, denoting those parts of the company's worldwide system under government control. This line just excluded the cable station buildings, leaving Porthcurno as the only station in Britain still under company control. The exiles were exiled. To quote J.E. Packer in *The Porthcurno Handbook*, 'In many ways Porthcurno now came to be regarded almost as if it were overseas.'

In 1951, 1,800 miles of cable were renewed in the Red Sea, and 1,400 miles of new telcothene-insulated cable was used to reinstate the Porthcurno–Halifax (Nova Scotia) telegraph. The 1952 Harbour Grace cable was the last cable to be laid by the company.

A positive and significant change for Porthcurno was the re-establishment of a Cable & Wireless Engineering College in 1950 for students destined for service overseas. Over the next 45 years the maintenance and management of integrated subsea cable, long-distance wireless and satellite networks became progressively more complex. In parallel, life still needed to be coaxed from legacy cable and radio systems, some of which had been installed many decades in the past. From now on this was to become the primary purpose of Porthcurno.

Cable & Wireless had also invested in an outside broadcast bus, which Porthcurno now owns and occasionally takes to classic car shows and heritage events in Cornwall. It is garaged at the Geevor Tin Mine. The bus had (and still has!) a narrow chassis to allow it to get through the gates at Lord's Cricket Ground; just as Marconi had made money out of betting (with the use of longwave radio to report on, for example, the Admiral's Cup), there was money to be made from reporting on sports events. The BBC were also providing TV feeds from these events, including monochrome transmissions, for example from a 1960 Humber Super Snipe with a Marconi Mark III camera, then colour from 1969, including some dashingly good-looking Citroën DS cars with a TV camera on the roof. These were replaced by outside broadcasting vehicles with satellite dishes.[57] These were always difficult to park, particularly in London, and have now been replaced with broadcasting backpacks with modems that use a multi-network SIM to set up parallel 4G and 5G connections.

The 1950s introduced new challenges and opportunities for subsea cable in general, as well as for the bus, Porthcurno and the Telecoms Coast. But the most important challenge and opportunity was long-distance subsea telephony.

The telegraph to telephony transition

In 1876 Alexander Graham Bell patented the telephone. It was exhibited at the Philadelphia Exhibition and more or less immediately started to be used by businessmen in Boston. The use of copper rather than galvanised iron made all the difference;[58] it made long-distance telephony not only possible but potentially profitable. In the UK William Preece (who reappears in Chapter 6) did his best to persuade the Post Office that telephony was the next big thing, but the Post Office failed to recognise that people enjoy talking to each other; the bean counters at the Treasury, already exasperated by the unprofitability of terrestrial telegraph networks, were unenthused at the prospect of funding a potential competitor. In addition, in 1880 the Liberal William Gladstone had campaigned to become Prime Minister based on reversing the imperialist foreign policy pursued by the Conservative Benjamin Disraeli. Two years later Gladstone was struggling with the unexpected cost of invading Egypt. Putting money into an untried device from America was low priority.

Meanwhile the privately owned Bell Telephone Company, the predecessor of the American Telephone and Telegraph Company (AT&T) was going gangbusters on the other side of the Atlantic. The first Strowger exchange burst into electromechanical life in 1892 in Indiana. The first Strowger exchange in Britain was installed in 1912 in Epsom, by which time America had over 7 million subscribers; the Post Office, with the help of Gladstone, had managed to put Britain 20 years behind in the race to develop a modern communications network. Making the GPO a state monopoly in 1912 didn't help much, either.[59]

In the meantime, Lee de Forest had been putting the finishing touches to his vacuum tube. Patented in 1907, the device was to transform the delivery economics of long-distance wireline and wireless voice communication. It took a while for valve technology to mature, but by 1915 the first long-distance telephone line between New York City and San Francisco was up and running. Initially it was only a single two-way circuit, but with the introduction of carrier multiplexing techniques from 1918 onwards, four or more voice channels could be transmitted simultaneously over two-wire or four-wire circuits.

By 1927 there were more than 3 million miles (5 million km) of long-distance telephone circuits in the United States. This traffic created an inexorable pent-up demand for transatlantic voice circuits, either over wireless networks or under the sea.

Telephones under the sea

On 24 December 1906, the year in which Marconi established a longwave telegraph service from Clifden on the west coast of Ireland to the east coast of America (via Newfoundland), Reginald Fessenden transmitted the world's first radio broadcast from Brant Rock in Massachusetts, using amplitude modulation (AM) to modulate the sound waves captured from a microphone on to a radio carrier wave. The era of radio broadcasting and radio telephony had arrived.

Progress in transmitting voice under the sea was slower. Cable quality had improved and inductive loading techniques had helped reduce the capacitive effects of salt water, so getting cables to carry voice circuits across short distances in shallow water was feasible – but long deep-water links were more challenging. By the late 1920s the consensus seemed to be that a long-distance cable supporting a single voice channel might be possible but would be significantly more expensive than doing the same thing over radio, particularly shortwave radio. A transatlantic cable was costed at $15 million. Given the Stock Market meltdown in October 1929, this was a proposal which (to mix metaphors) was never going to fly.

The first commercial voice link, a single radio telephone circuit, was established across the Atlantic in 1927. Being shortwave, it was not perfect. It usually worked better at night (what was known as dark path routeing), and the noise floor increased as the sun became more active over the 11-year sun spot cycle, but it worked adequately well most of the time. Radio was also seen as a more insecure communications option in comparison to cable – though, as we shall see in Chapter 8, this was not entirely the case.

Valves change the delivery economics of subsea, terrestrial and radio networks

As smaller, more efficient valves became available in the 1930s – valves were needed to amplify the signal – it became increasingly possible to consider a North Atlantic submarine telephone cable, but the valves were initially considered to be too fragile. By the Second World War, however, valves were being used in a range of harsh environments, and there was a substantial investment in making the devices robust and reliable. Much of this work was done in the Dollis Hill Research Labs in North London. By the 1940s the Post Office had developed a single repeater system which was deployed in shallow water. In 1950 the first prototype of what was to become known as the Trans-Atlantic Telephone (TAT) Cable was deployed between Havana and Key West, capable of meeting the demands of being deployed at depths of over 2 nautical miles.

The first iteration of this cable, announced in 1953 and known unsurprisingly as TAT-1, was a joint venture between AT&T Bell Laboratories, the British Post Office Engineering Department and the Canadian Overseas Telecommunications Corporation. The repeater was flexible enough to be wound over a standard cable drum. It was 8 feet long and had a diameter of 2.875 inches, tapering down to the cable width of 1.625 inches over a length of 20 feet.

The main Atlantic link, designed by the Bell System company, had two cables, one for each direction, with one-way flexible repeaters at 37-mile intervals. HMTS *Monarch*, the world's largest cable ship at the time, laid the two cables in the summers of 1955 and 1956. The Canadian launch point was Clarenville in Newfoundland. Disappointingly for Porthcurno, the other end of the cable landed at Oban in Scotland, but, in a way, this was inevitable; land links to the north of Scotland were now much improved (backed up by VHF point-to-point radio links) so Oban was a logical choice. The Canadians also had an overland line-of-sight radio link from Nova Scotia to Montreal.

Each of the two cables had 51 repeaters, spanning 1,950 nautical miles. They provided 65 dB of gain and 144 kHz bandwidth at a centre frequency of 164 kHz with signal amplification realised with three vacuum tubes.[60] These valves worked faultlessly for 22 years, supporting twenty-nine telephone circuits between London and New York, six circuits between London and Montreal, and a single circuit split among the three destinations for telegraph and other narrow-band applications.

Transistors turbocharge terrestrial and subsea copper networks

As with the terrestrial telephone network, improvements in transistor technology from 1950 through to the present day (the doubling of device density every 18 months, in accordance with Moore's Law) transformed subsea cable system performance, as did digital switching and the introduction of digital codecs and advanced time-division multiplexing in the 1970s.

Optical cables arrive in Porthcurno

In parallel, from 1970 onwards, year-on-year improvements in optical fibre technology gradually edged copper out of terrestrial and subsea networks. This was due partly to throughput and partly to costs, including the maintenance cost; optical fibre has lower maintenance overheads than copper. Appendix 2 has more detail on optical subsea technology from 1970 through to the present.

On 31 December 1970 after a cost review by PA Management Consultants, Porthcurno closed as a cable station. The engineering college closed in 1993.

However, new cables still arrived on the beach. On 13 May 1997 the Gemini transatlantic cable was the first transatlantic optical cable and was capable of carrying 1.5 Gbps of data. An equivalent cable today carries 250 terabits per second, but 1.5 gigabits was impressive at the time.

Porthcurno – or more specifically Cable & Wireless, and subsequently Vodafone – was responsible for the FLAG (Fibre-optic Link Around the Globe) cable, though the transition from subsea cable to terrestrial cable was close to the Sennen Cove End at Skewjack Farm. The significance of this is covered in Chapter 8.

FLAG is a 28,000-kilometre cable connecting 18 countries and regions in Asia, Africa and Europe. It opened for commercial service on 2 November 1997. The cable was acquired by Reliance Communications in 2003 for $207 million.

The cable was routed from Porthcurno to Estepona, near Malaga, in Spain, Palermo in Sicily, Alexandria and Cairo, Suez, Aqaba in Jordan, Jeddah in Makkah Province (Saudi Arabia), the United Arab Emirates, Mumbai, Penang in Malaysia, Thailand, Hong Kong, Shanghai, South Korea and Japan.

Fibre subsea cables still have a conductor (glass, not copper), an insulator (polyurethane rather than gutta percha) and a strength member (steel instead of iron) but have a substantially tighter bending radius than copper cable and can be deployed from relatively small ships (450 feet rather than the *Great Eastern*'s 700 feet) at a laying rate of 150 kilometres per day. Technology moves forward on multiple fronts.

Optical subsea routeing in 2025

As of 2025, the optical trade routes (just like their subsea copper predecessors) are still dominated by traffic from the West into the USA, and then from the west coast of the USA across the Pacific Ocean. (The Washington-based consultancy Telegeography maintains a continuously updated map of these routes). Essentially, the UK and the Cornish Telecoms Coast is still an important part of this (in modern terms) subsea global delivery ecosystem.

A shift towards the Equator

Global routes are unlikely to change in the immediate future, but there is a discernible shift towards subsea routes that run closer to the Equator. These new optical routes complement space-based optical and radio backhaul connections, including the integrated switching of traffic between geostationary satellites and Medium and Low Earth Orbit satellite networks, a topic for Chapter 7.

A precursor of this shift is Meta's newly announced cable project, Project Waterworth. This is a 50,000-kilometre subsea cable budgeted to cost $10 billion. When completed, this will be the world's longest optical cable, and it will be running at depths of up to 7,000 metres (7 km).

To quote the press release:

Subsea cables projects, such as Project Waterworth, are the backbone of global digital infrastructure, accounting for more than 95% of intercontinental traffic across the world's oceans to seamlessly enable digital communication, video experiences, online transactions, and more. Project Waterworth will be a multi-billion-dollar, multi-year investment to strengthen the scale and reliability of the world's digital highways by opening three new oceanic corridors with the abundant, high-speed connectivity needed to drive AI (Artificial Intelligence) innovation around the world.

This is a 24-fibre pair cable capable of carrying between 250 and 500 terabits per second.

A shift in cable ownership

As of 2025 the world is interconnected by 1.3 million kilometres (800,000 miles) of subsea cable. None of these cables are owned by Cable & Wireless, as it no longer exists as an independent company. Instead, they are owned by companies that did not exist 50 years ago (Microsoft) or 30 years ago (Google, Amazon, Facebook/Meta, Apple, TikTok, Tencent and BeiDou).

In February 2021 Google and its engineering sub-contractor SubCom commissioned the Dunant subsea cable connecting Virginia Beach in the United States to St Hilaire de Riez on the French Atlantic coast. Named after the founder of the Red Cross, the 6,600-kilometre cable carries 12 fibre pairs and is claimed to be the first long-distance cable to use space division multiplexing (SDM), effectively multi-mode fibre. (See Appendix 2 for an explanation of this bit of technobabble.)

Each fibre pair supports a data rate of 25 Tbps yielding a cable capacity of 250 Tbps. The use of SDM (see Appendix 2 again) means that pump lasers and related photonic and optical components can be shared across all fibre pairs opening the opportunity to realise higher pair counts, up to 24 pairs being regarded as practical. As of today 16 fibre pairs can carry 180 Tbps over subsea paths of thousands of kilometres, with visibility to 500 Tbps in the near future.

Today this technology is making money for companies who have a collective market value of over $20 trillion.

In 1981 Cable & Wireless was a bigger company than BP (British Petroleum), employing 54,000 working all around the world. It was one of the only remaining founders of the original FTSE 30 stock exchange. By the late 1990s the members of the Board were congratulating themselves on the success of their reinvention of the company as an Internet services business, a business they knew little about.

At the start of 2000 the C&W Group was worth £38 billion, and the Board members could sit and admire some fine newly commissioned artwork (the pictures are still stored in the archive room at the Porthcurno Museum). Never put money into a company that invests in china teacups and art; by 2015 Cable & Wireless no longer existed.

How did Cable & Wireless miss out on this technology-driven renaissance of the subsea cable industry, and how did it manage to destroy the best part of $40 billion of shareholder value in just over 15 years?

The potted history from fame and fortune to failure

The rise and fall of Cable & Wireless can be explained by summarising ten key dates, marking either a pivotal moment in the onward march of telecommunications technology and telecommunications and networking standards, or a parallel shift in regulatory and competition policy.

Ten steps to oblivion in date order

1979: The first standards for the Internet.[61]
1981: Cable & Wireless is privatised under Margaret Thatcher with the help of David Young, later Lord Young of Graffham. Mercury Communications is set up to compete with BT (a politically driven rather than commercially driven decision).
1987: The handover of Hong Kong to China (for many years Hong Kong had been the main source of profit for Cable & Wireless).

1989: The first standards emerge for the World Wide Web.
2000: The NASDAQ index peaks at 5,048 on 10 March 2000. By October 2002 the index had lost 76.81 per cent of its value. Many dot-com stocks go into Chapter 11, joining Iridium, Globalstar and Teledesic (the first generation of LEO space radio networks; see Chapter 7).
2002: The world's first camera phone, the Nokia N650. Twenty years later, sub-metre resolution from space is transforming the economics of the satellite industry, and it all started in Finland.
2007: The first-generation Apple iPhone.
2009: The decision to set up two separate businesses, Cable & Wireless Communications and Cable & Wireless Worldwide.
2012: Vodafone acquires the wireless assets of the group (Cable & Wireless Worldwide) for £1 billion.
2015: The rest of the business is sold to Liberty Global for £3.5 billion, including £1.8 billion of debt. In 2023 Liberty Global made a loss of $3.9 billion and had a debt of $15.9 billion.

By 2010 the denizens of the hedge fund world were gunning for the dismemberment of what was left of the Cable & Wireless Empire.

The problem that Cable & Wireless had is like that of other UK privatised utilities, including most recently the water industry. Utilities are attractive to private investors because they are considered to have captive customers yielding stable revenue and high margins. However, this also makes the companies easy to load with debt, and this makes them sensitive to price erosion or loss of revenue from new competition, particularly competitors with large amounts of cash on their balance sheet generating what is called Over the Top Value (OTT) from networks which provide bandwidth to them at a discounted price.

As this new competition begins to bite, hedge funds start shorting the stock and the debt load becomes more expensive to service. For Cable & Wireless, privatisation came hand in hand with market liberalisation. This eradicated the margin premium achievable from market monopoly.

The bursting of the dot-com bubble in March 2000 and the oversupply of fibre bandwidth made things even harder. In the late 1990s Cable & Wireless moved into data hosting and invested in US companies that were overvalued. As with GEC Marconi, it could be argued that the decision to buy into a US supply chain and a new market (the provision of Internet services and data centre hosting) was the right decision at the wrong time.

What happened next can be regarded as nobody's fault, and the fate of Cable & Wireless was shared with many other British companies that prospered in the 1990s then crashed and burnt over the next ten years (GEC Marconi being a notable example, a topic for our next chapter). The shareholders and stakeholders of both companies were, however, less than pleased.

With the benefit of hindsight, if Cable & Wireless had remained faithful to its core business (no pun intended) of cable laying, cable termination and cable repair, and if the company had continued to invest in its wireless and space businesses (to be covered in the next two chapters), it could now be providing integrated terrestrial, subsea and RF and optical space communication as a trusted delivery partner to most of the rest of the world.

The Telecoms Coast: from low ebb to high tide

By 2015 Cable & Wireless no longer existed. In 2008 the telecom activities at Goonhilly were transferred to Madley in Gloucestershire, and the Dishes on the Downs seemed to be destined to join the pumping engines of Cornwall, providing evidence of a glorious but distant industrial past. The Marconi Company was bankrupt, having sold the last of its assets at a knock-down price to Ericsson.

The Telecoms Coast in Cornwall was, to all appearances, dead and buried.

In reality, the Telecoms Coast is alive and well and connecting Britain to the rest of the world via a combination of cutting-edge RF, optical terrestrial and space communication technology. So bear with us; this story has a happy ending.

Waves rise and fall and rise again. It is time to turn our attention to the master of the radio wave, Senatore Guglielmo Marconi.

6

STEAM RADIO TO BEAM RADIO

In Chapter 5 we left Guglielmo Marconi in 1927 at a low point, forced to move his long-distance wireless assets into the new company that would become Cable & Wireless. The first of many minor – and, later, not-so-minor – heart attacks did not help, either. He had separated from his first wife, Beatrice O'Brien (who had found a new Italian husband), and married Christina Bezzi Scali, who was younger with aristocratic Italian connections (similarly, Beatrice had aristocratic Irish connections). Marconi still had a manufacturing business to run, which he still more or less owned, an annual research budget from Cable & Wireless, and a fancy suite of offices in Victoria, where he didn't spend much time.

He was to die in Rome ten years later, on 20 July 1937 aged 63, but as we shall see his legacy lives on; the pioneering work that his engineering team did on what came to be known as beam radio remains relevant to the new wave of radio and space technology being deployed today, just a few miles from where his Cornish adventure started on the Lizard Peninsula.

This chapter starts with the story of Marconi's first four years in London, from 1896 to 1900, and the creation of a maritime radio business, which began in the Isle of Wight and worked its way down to the Lizard and then to Poldhu Cove. The maritime business initially made relatively small amounts of money, but it is a curiosity of capitalism that you do not need to make money to make money; between 1898 and 1900 there was a six fold increase in the value of the stock of the Marconi Wireless Telegraph Company. The company was still several years away from making a profit, and didn't pay a dividend until ten years later, but the stock appreciation was sufficient to float the American Marconi company with a capitalisation of $10 million. As the business was defined as being based at sea (outside the 3-mile territorial limit), the Post Office terrestrial monopoly over telegraphic services did not apply. (The Post Office had been recently reorganised as the GPO, with William Preece as Chief Engineer).

The maritime business was a means to the end, the end being the end of the subsea cable business.

By 1900 Marconi had proved to himself and to his investors that, contrary to the general scientific consensus at the time, radio waves do not necessarily travel in straight lines but can follow the curvature of the Earth, achieved by a combination of what we now term as ground wave and sky wave propagation, in which the Earth's atmosphere acts as a wave guide. This meant that given sufficient transmit power and a sufficiently sensitive receiver, messages could be sent by wireless around the world at a fraction of the cost of a terrestrial or a subsea cable. As there were no intermediate repeater stations, radio was also at least an order of magnitude faster.

Up to 1900 nearly all wireless equipment had been battery-powered. Terrestrial and subsea copper networks were also low voltage. The copper networks presently being decommissioned around the world today still work on a 12-volt supply.

Marconi realised that long-distance radio – defined as a radio system that could cross oceans and continents – would need thousands of volts to create the energy needed to send a radio wave thousands of miles. This required energy to be produced efficiently and used efficiently. An engine (oil, diesel, or steam), or hydroelectricity, efficiently coupled to an antenna system designed to coexist comfortably with other geographically proximate radio systems.

The story goes as follows:

In 1896 Marconi arrived in London with his mother. He set up a maritime radio business which turned out to be the start of a much bigger enterprise.

In 1898 he opened the world's first purpose-built radio factory in Chelmsford.

In 1899 he recruited Ambrose Fleming, initially as consultant then as a full-time member of a remarkable team of engineers. Fleming had worked with Edison, so he understood how to produce AC and DC power efficiently and more or less safely.

Marconi then built a succession of long-distance radio stations, starting with one at the Lizard, which was the prototype for the more powerful radio station at Poldhu, which was the prototype of an even more powerful (steam-driven) radio station at Clifden on the west coast of Ireland,[62] which was the prototype of an even more powerful (hydro-electric powered) radio station at Waunfawr in Snowdonia. All of these were longwave; Poldhu was initially 819 kHz (366-metre wavelength) then 272 kHz (1.1-kilometre wavelength) then 70 kHz (4.2-kilometre wavelength), Clifden was 62.4 kHz (4.8-kilometre) then 42–45 kHz (6.666-kilometre) and Waunfawr was 21 kHz (14-kilometre). In 1905 Fleming developed a way of measuring wavelength, removing some of the guesswork from the system design process, though the team were still in the dark on some of the fundamentals of longwave radio propagation.

All we need to know for the moment, though, is that longer (lower) frequencies need longer antennas and therefore take up more space. Compared to anything else being built at the time, Marconi's radio stations from Poldhu onwards were huge (Clifden covered 300 acres and had the largest battery ever built at the time) but energy-inefficient; the peat bog feeding the six steam engines at Clifden disappeared at an alarming rate. However, Charles Franklin, one of Marconi's growing team of young engineers, had been leading a team based at Poldhu developing shortwave beam radio, and by the 1920s the world had an efficient global wireless radio network. From steam radio to beam radio in just over 20 years!

This is the summary of a story that deserves to be told in more detail.

Marconi: following in the footsteps of Brunel and treading on the toes of the (later) Penders

The subtitle for Chapter 2 was 'How Brunel Connected Cornwall, How Cornwall Connected the World' (via the Penders and Porthcurno). An equally apt subtitle for this chapter would be 'How Marconi Connected Cornwall, How Cornwall Connected the World (via Mr Fleming and Poldhu)'.

Marconi started life well connected, remained well connected all his life, and died well connected. He knew both Roosevelts – Teddy and Franklin D. – and was a friend of the Prince of Wales (the future George V) and Sir Thomas Lipton (the tea magnate and ocean racing enthusiast). If you were asked to define Marconi in two words you could say 'quietly charismatic'.

On the other side of the Atlantic, Marconi also rubbed shoulders with the magnates of the day including JP Morgan (banking), Andrew Carnegie (steel), John D Rockefeller (oil) and the Vanderbilts (railroads and shipping). This was America's gilded age, and Marconi somehow managed to be part of it while simultaneously creating a global wireless network.[63]

In 1909 he was awarded the Nobel Prize in Physics, which he shared with Karl Ferdinand Braun for 'their contributions to the development of wireless telegraphy'. Ambrose Fleming was on the nomination committee. Then in 1910 Marconi lost his right eye driving too fast in Italy in one of the many fast and splendid cars that he owned through his life. The replacement glass eye was so well painted that few people realised it was false.

After serving in the Italian navy in the First World War (researching shortwave radio and radio direction finding) he befriended two Popes and developed a complicated relationship with Mussolini (Marconi's boyhood friend Margherita Sarfatti was one of Mussolini's mistresses). He was appointed a senator in 1914 by the King of Italy, Vittorio Emmanuel III, and later became a marchese. Perfectly fluent in Italian and English, Marconi was quietly spoken, and persuasive and effective when defending one of his many patent disputes in court. He was a talented pianist. His elder brother Alfonso was an equally talented violin player.

Marconi's mother and Hertz

Marconi was born on 25 April 1874 in Bologna. His mother, Annie Jameson, a scion of the Jameson whiskey family, had met Marconi's father, Giuseppe, when studying singing at the Bologna Conservatoire. Giuseppe, 15 years older than Annie, was a successful industrialist with a country estate, the Villa Griffone, just outside Bologna. Marconi is buried on the estate in a mausoleum built by Mussolini in his honour.

Marconi spent two years, between the ages of 5 and 7, at school in England. At 13 he went to the Leghorn National Institute, attended Augusto Righi's lectures on electromagnetism, and studied Heinrich Hertz, at which point he became fascinated by radio waves.

Hertz was born in 1857,[64] and between 1886 and 1888 had carried out experiments at around 250 MHz (remarkable in itself, as these VHF frequencies only became widely used in the 1940s) to prove Maxwell's theories of electromagnetic properties, by showing how an oscillatory spark discharge could produce a heavily damped radio wave from an initial spark across an air gap, followed by a number of sparks reducing in intensity until they were too weak to bridge the gap.[65]

In 1890 the French physicist and inventor Édouard Branly had produced a device which came to be known as a coherer. This was a glass tube containing iron filings which coalesced in the presence of a radio wave, at which point the filings became conductive and passed current to a bell or a light, a radio receiver. In the spring of 1895 Marconi built his first spark transmitter upstairs in the villa; then he crafted a receiver by remoulding a thermometer, using a hand bellows to create a vacuum in the tube, adding iron filings and a small hammer, which tapped the filings to shake them apart ready to detect the next transmission. He enlisted his older brother to help with a range of experimental radio transmissions in and around the estate, which involved coupling the spark transmitter to lengths of wire of various lengths earthed to ground and convinced his mother and father that he was going to conquer the world with his new 'transceiver device', which he built into a nice wooden box.

He arranged to show the box of tricks (because this is what it came to be called) to the Italian Ministry of Posts and Telegraphs, who were unsure what it could or should be used for – though it was suggested (and the evidence here is somewhat vague) that it might be of use for maritime ship-to-ship and ship-to-shore communication. With the benefit of hindsight, it seems blindingly obvious that the one thing that radio could do that cable could *not* do is connect things that move, including ships. If connecting ships meant that fewer ships would sink, then that would be a good thing, and if connecting ships meant that when ships were sinking, passengers and crew could be rescued then that would be a good thing too; and if naval ships could be commanded by radio then that would be another good thing – and if telegraph services and news services could be delivered to transatlantic steamships and to and from swanky sailing boats, then that would also be a good thing.

Ironically, however, Guglielmo had failed his exams for naval college, so instead in February 1896 he travelled to London, still the world capital of shipping (though with America catching up fast). Crucially, the Post Office monopoly did not extend to maritime communication. Annie Jameson introduced Guglielmo to her nephew, Henry Jameson Davis. Twenty years older than Marconi, Jameson Davis was a milling engineer (the essence of whiskey making) with well-connected wealthy friends and an expert knowledge of the patents process. William Preece, a friend of a friend, was sufficiently impressed to introduce the device to a sceptical Post Office (at this point William Preece was 62 and Chief Engineer of the GPO, whereas Marconi was 22). It turned out that the Navy were interested, and so was Trinity House. Jameson Davis raised £100,000 from his corn merchant chums, allocating Marconi £60,000 in shares with £25,000 for research and £15,000 for patents. The Wireless Telegraph and Signal Company was set up in July 1897.[66]

The patents

On 2 June 1896 Marconi filed his first patent (Number 12039) for 'a sensitive tube receiver, or coherer, connected to an earth or elevated aerial and the timing of the transmitting circuits with each other'. Over the next 40 years, 800 patents were granted to Marconi and Marconi

Companies (the company in its various guises continued to patent new radio system technologies through to the 1990s). One of the most famous (and infamous) was the Four Sevens Patent, named after its issue number (7777), issued on 26 April 1901.

The patent was 'inspired' by an experiment patented in 1897 by Professor Oliver Lodge, covering the tuning of the sending and receiving aerials. Marconi's patent was for selective tuning in which other circuits in the transmitter and receiver, as well as the aerials, were tuned to the same wavelength. An adjustable inductor is used in the antenna feed; the transformer is tuned on both the primary and the secondary windings, and there are variable inductors in series with the split secondary windings. Two iron-cored chokes block RF from the relay or 'telegraph instrument'. The main tuning is by two variable capacitors constructed from two metallic tubes separated by a dialectic. The tubes slide telescopically over each other. This was elegant, though not uniquely novel, at the time.

Apart from power, radio systems need sensitivity on the receive path, selectivity in and between the transmitter and receiver and frequency stability. In 1902 the Marconi Wireless Telegraph Company patented the Magnetic Detector, generally known as 'Maggie'. This was a clockwork device moving a soft iron band at a steady 8 centimetres per second through a tube and past permanent magnets, with a primary winding connected to the aerial and a secondary winding connected to a pair of headphones. When a signal was received from a spark transmitter in the primary winding, it changed the magnetic field induced in the iron band as it entered the tube. The variation of magnetism induced a current in the secondary winding and produced clicks in the headphones. This device was more sensitive than the coherer and became the standard detector used in ship and shore radio systems until replaced by crystal detectors and then valves. Some spark transmitters were still in use in the 1920s, though by that time they were producing embarrassing amounts of interference into other radio systems. Valves were more efficient at detecting radio signals and, along with the use of simpler crystal detectors, were the only way to recover voice modulated on to a carrier wave.

Maritime Marconi

Marconi loved being on the sea or by the sea and was fond of comfortable hotels. By November 1897 he had built a 120-foot aerial close to the Needles and Alum Bay and started sending radio signals to coastal steamers in the Solent. In January 1898 a second wireless system was installed, in the Madeira Hotel in Bournemouth. Marconi then moved to the Haven Hotel in Poole. By 1899 messages were being sent across the Channel and Marconi made a splash at the Cowes Regatta, setting up a radio relay between Queen Victoria at Osborne House and the Prince of Wales. In October he was in New York providing radio links for Gordon Bennett – yes, *that* Gordon Bennett – reporting on the America's Cup sailing race. This was valuable publicity, and a proving ground for working out how to get further by radio. Putting antennas on ships produced one of the first 'rules of radio'; doubling the height of the aerial on the mast meant that a message would go four times as far. Marconi received $5,000 for his efforts in New York and generally had a nice time, but he was spending money faster than he was making it and something needed to change. The technology was taking shape, but the company needed engineering expertise. It is time to introduce Ambrose Fleming and Charles Franklin.

Fleming (1849–1945) and Franklin (1879–1964)

Ambrose Fleming had studied with James Clerk Maxwell (1831–1879) and worked for Thomas Edison (1847–1931) through the 'wars of the currents' (the DC versus AC debate). Fleming's light bulb moment came in 1904, when he worked out that Edison's light bulb could be repurposed as a diode valve – but we are getting ahead of ourselves. In 1899 Marconi employed Fleming as a consultant. In December 1900 Fleming became a scientific adviser and worked for Marconi for the next 30 years. As we shall see, he became master of the art of turning miniature bolts of lightning into a global telecommunication system.

Also in 1899 the 20-year-old Charles Franklin joined the Marconi Company and was sent to South Africa with radio kit for the Boer War (which didn't work well, as no one knew how to use it). He spent two years in Russia, then came back to invent the variable capacitor (patented in 1902), ganged tuning (multiple tuning stages, patented in 1907), variable coupling (also patented in 1907), coaxial cable, an oscillator which came to be known as the Franklin oscillator, and his crowning achievement (largely developed in Poldhu), the Franklin Beam Aerial.

Beam forming is at the heart of all modern radio systems. It is used on the 4G and 5G cell sites that your smartphone talks to all day, and is the critical enabling technique that allows satellite earth stations (including Goonhilly) to talk to thousands of Low Earth Orbit satellites tracking from horizon to horizon at 17,000 miles per hour.

In simple terms, if you take two collinear antennas (two vertical metal poles) and move them apart, the two antennas move in and out of phase with one another, either reinforcing the waves of signal energy from the antennas or cancelling the waves of signal energy. The result is that you can create signal nulls where you don't want the signal to go, or signal nulls in the direction of unwanted signal energy from interfering radio systems. This reduces the amount of received interference. You can also take whatever radio energy you are coupling into the antenna elements (in this case our two antennas) and send the energy in a particular direction, for example Australia or South Africa. This increases the amount of signal energy sent to where it is needed. In a modern 'smart antenna' you could have for example 1,024 antenna elements in a flat panel phased array antenna, which can sweep fractional degree signal lobes around from horizon to horizon at millisecond time scales. Franklin would be impressed, but probably also pleased that his pioneering work would prove to be fundamental to contemporary radio economics.

Some basic directivity was possible at longwave[67] but required acres of real estate (hence the move to Clifden, of which more later). At shortwave, beam forming helped transmission efficiency (The benefits of beam forming increase with frequency). Present 5G systems operate at up to 4 GHz and may move to 28 GHz and 32 GHz. Satellite systems operate in the same bands and above (Appendix 2).

Although Poldhu is most famous for the 'Atlantic Leap' in December 1901, the first shortwave transmissions to Marconi on his yacht *Elettra* in the South Atlantic during 1923 and 1924 are at least as important historically. Later in his career (in 1935) Franklin designed the VHF antenna at Alexandra Palace for the first-generation 'high definition' (405-line) TV broadcasts. By the 1930s the Marconi company was a significant manufacturer of radio and TV sets and radio and TV broadcasting equipment including high-powered radio valves, though that is a chapter in another story. Back in 1924 Patent No. 242342 for 'a pronounced directional effect from aerials of the type that are electrically long in comparison with their signal wavelength' marked a significant step forward in radio system design and radio system efficiency.

A memorial on the headland at Poldhu marks the work of Charles Franklin on beam radio with the following inscription:

> From the Marconi Company's Poldhu Station in 1923 and 1924, Charles Samuel H. Franklin, inventor of the Franklin Beam aerial, directed his short-wave wireless beam transmission to Guglielmo Marconi on his yacht *Elettra* cruising in the South Atlantic. The epoch-making results of these experiments laid the foundation of modern high speed radiotelegraphic communication to and from all quarters of the globe.

Meeting more of the technical and engineering team

Fleming and Franklin were part of a small but dedicated and hardworking technology and engineering team including George Kemp, who generally looked after the building and repair of aerial and mast systems (and usefully kept a detailed diary), Percy Paget (who seemed to be able to turn his hand to most things) and Richard Vyvyan, who usefully produced a book in 1933 recording his 30 years of working with Marconi.[68]

It is worth quoting from the book as a precursor of our tour of the high-power radio sites. Writing about the early years when the Lizard and Poldhu radio stations were being developed. He comments:

> Marconi had shown that increasing the height of the aerial increased the range of signalling in proportion to the square of the height, the practicable economic height of wooden masts for supporting the aerial was thought to be about 200 feet.

(It was not thought wise to use steel masts or towers until a much later date presumably due to the risk of lightning strikes and electrocution.)

The conclusion he arrived at was that transatlantic wireless could only be accomplished by the radiation of much greater energy than could be obtained from an induction coil and Leyden jars and that therefore the transmitter would have to be designed as an engineering plant of considerable power yet arranged so as to be safe to use.

The Wireless Company in 1899 had appointed Professor J A Fleming, FRS, of University College, London, as Scientific Adviser to the Company,… knowing the experience that Dr Fleming had gained in dealing with extra high tension alternating currents in electric lighting work … they decided that it would be necessary to employ an alternator driven by an engine and actuating high tension transformers in order to charge a bank of condensers which would discharge through an oscillation transformer across a spark gap and they further agreed that the alternator should have an output of 25 kilowatts.

(They also needed] to devise some method of interrupting the primary circuit of the transformers to allow signalling at will [preferably without killing the operator).

It was decided that the station for the transatlantic experiments should be built on the west coast of Cornwall in some isolated place, in order to avoid the possibility that the powerful electrical oscillations might affect electric circuits used for lighting or other purposes in the neighbouring district, and also in order that there should be no obstacle to the clear passage of these waves over the ocean in the immediate vicinity of the station.

Richard Vyvyan had experience working on power stations, so was given the job of assistant to Fleming, helping to design the plant and to order and/or arrange the manufacture of the high-tension equipment and power condensers. In July 1900 Vyvyan went with Marconi and the new Managing Director of Marconi Wireless, Major Flood Page (who had taken over from Henry Jameson Davis) to find a suitable site, which turned out to be Poldhu. The site became available in October; Richard Vyvyan started work on the equipment and George Kemp started work on building the aerials.

It must have been exciting work, and Marconi would have paid well – but there must also have been a feeling of common purpose. By all accounts Marconi could also be a convivial companion. He also managed to get Thomas Edison actively involved in the company. Edison was always passionately interested in telegraphy, his two eldest children were nicknamed Dot and Dash. In 1903 Edison joined the Marconi Board and in 1905 swapped his wireless patents for a stake in Marconi America.

Meanwhile maritime keeps the Marconi Company afloat

The Marconi Company would never have survived without the perceived future promise of profits from the maritime business. The Marconi Maritime Telegraph Company had been set up in London in 1899, and Trinity House and Lloyd's were early and continuing customers. The business was basically putting equipment (made in Chelmsford) onto merchant ships, naval ships (though they were slow adopters) and transatlantic liners.

Operators were trained by the Marconi Company and ship owners were charged for the service. The original plan was to have eight coastal stations around the UK, including St Catherine's Point on the Isle of Wight, the Lizard and Land's End. By March 1902 there were 25 wireless stations. On 26 September 1902 Marconi Marine signed a 14-year agreement with Lloyd's.

By 1912 all transatlantic liners had a wireless telegraphy cabin with a 200-mile range spark transmitter for distress calls and message services for first class passengers.

The sinking of the *Titanic* on its inaugural voyage on 15 April 1912 is the highest-profile example of the use of a distress call to save lives (Marconi and Beatrice were both booked on to the voyage but had to cancel at the last minute) but the Lizard Radio Station and Land's End regularly fielded distress calls. These included one in April 1910 from the *Minnehaha*, which had run aground east of the Scillies on Seal Rock near Bryher, picked up by the Lizard Station, and from the 7,000-ton *Gothland* which had hit the rocks off the Scillies in thick fog in June 1914, picked up by Land's End.

The *Titanic* disaster prompted an International Conference on Safety at Sea in London. On 20 January 1914, 16 nations agreed on 74 Articles for improved safety in shipping, including the use of wireless. Well over 3,000 British ships were sunk in the two world wars. Significant losses in the First World War include the sinking of the *Lusitania* in 1915 (which brought America into the war); the liner, torpedoed off the coast of Ireland, sank in 18 minutes. Marconi knew many of the staff and crew. From 1919 all passenger and cargo ships over 16,000 tons had to have wireless systems and an operator.

Marconi crossed the Atlantic at least 40 times, so had a vested interest in what is generally termed Safety of Life at Sea (SOLAS) and always made sure to spend time with the telegraph operators (normally two operators working shifts on the larger ships). Most of the day-to-day value was handling telegraph messages and providing passengers with daily news bulletins and stock updates, but any big disaster at sea always translated to an increase in the Marconi share price. By the 1930s over 3,000 ships carried Marconi wireless systems.

Transatlantic liners as a test bed

As more ships became equipped with Marconi radios, more testing could be done on radio coverage and optimisation. Richard Vyvyan writes:

> In February 1902, Marconi thought it desirable to test how far messages transmitted by the powerful station at Poldhu could be detected on a ship. The ship selected was the SS *Philadelphia* of the America Line and the results obtained during the voyage were of great scientific importance. The receiving aerial was fixed to the top of the mast which was about 170 foot high. As the aerial was fixed and not attached

to a kite floating up and down as in the Newfoundland experiment (a reference to the Atlantic Leap the year before), a syntonic receiver (a receiver based on the Four Sevens patent) could be used and the signals recorded on tape.

Marconi sailed on this ship towards the end of February. On the start of the voyage the ship was in communication with the ordinary short range coastal station on the Isle of Wight up to a range of 70 miles, after which Poldhu was tuned in. Readable messages were recorded on tape up to a distance of 1551 miles from Poldhu, and signals, the famous letter S (also used in the Transatlantic Leap) were recorded up to 2099 miles.[69] Tests were also made using the self-restoring Italian Navy coherer, as in the Newfoundland experiments, but it is of interest that with this coherer, 700 miles was the limit of distance.

The most important scientific discovery made on the voyage, however, was that electric waves as then used (Long Wave) travel further at night than at day. At distances of over 700 miles signals transmitted in daylight failed entirely, while at night they were quite strong at 1550 miles. Had Marconi known of this phenomenon when in Newfoundland and carried out night tests the results he would have achieved would have undoubtedly been more convincing.[70]

The voyage proved that with a similar station to Poldhu in service on the American Coast, shipping could be kept in touch with the shore across the Atlantic. On arrival in New York Marconi stated that given adequate apparatus, it would be possible to send a message entirely around the world, to start the message eastward around the globe and to receive it again at the same station from the westward. In 1926 this prophecy was fulfilled when the short wave beam stations were connected and I myself saw signals sent from one of the Imperial Beam Stations recorded not only once but three times at intervals between each record of one seventh of a second, proving that the signals had made a complete circuit of the world three times and recorded themselves on each circular tour'.

The maritime business kept the Marconi business afloat but also proved that a global wireless system was practical and possible.

Going global – from longwave to shortwave (power-efficient) radio

To go back to where we started, the Lizard was a prototype for Poldhu, which was a prototype for Clifden on the west coast of Ireland, which was a prototype for Waunfawr in Snowdonia. All of these were longwave and relatively power-inefficient. The development of a shortwave radio system on the headland in Poldhu paved the way for the move to an antenna system with directional gain, which provided the enabling technology for Bodmin (the first fully operational shortwave beam radio station), which was the first of the Imperial Wireless shortwave beam radio stations.

The Lizard Wireless Station was the first to be powered by an oil engine and generator rather than batteries (hence the description 'high-powered'). In August 1900 Major Flood Page and Marconi had travelled down to the Housel Bay Hotel, just south-east of the Lizard village. Half a mile to the east, the Lloyd's signal station provided signalling to ships by running flags up a mast. Flood Page approached Viscount Clifden, the local landowner, and arranged to lease a plot of land in the wheat field next to the hotel. The local builder, George and Sons, was employed to dismantle the Great Western Railway hut in the village. The building was 33 feet by 11 feet and had four rooms. (See Figure 3.1, Chapter 3, The Lizard Wireless Station, formerly the GWR Waiting Room.) A wooden mast, brought from Dovercourt in Essex in three 60-foot sections, was used to build a 150-foot mast supported by guy wires.

The transmitter was a ten-inch (25 cm) induction coil with an interrupter to provide a spark. Capacitors and a transformer were used for tuning before the signal was fed into the aerial. The capacitors were Leyden jars, large cylindrical jars coated inside and outside with foil, with the inner layer connected to a metal rod with the glass acting as a dialectic. Six jars were placed in a wooden tray, with the outer layers connected by a brass strip initially connected in parallel to the aerial. Some primitive tuning was achieved using a transformer on the aerial. The secondary coil had multiple tappings to change the turns ratio. The receiver was a standard coherer operating a relay which operated a Morse inker, a paper tape pulled through an inker (a pen) by clockwork. Earthing was achieved by having ten steel plates each 6 feet by 3 feet set into the ground around the hut. A 1.25 horsepower oil engine drove a 500-watt dynamo in one room, feeding the accumulator in the adjoining room. The battery consisted of 16 chloride cells.

Marconi arrived on 23 January 1901, having arranged that the station at St Catherine's Point on the Isle of Wight would send Morse signals that day – which were successfully received. It was 195 miles (315 km) between the two stations, confirming that over-the-horizon transmissions were practical and realisable. Marconi referred to this as his 'First Little Miracle'. Signals were also received from other coastal stations and from northern France. The daytime range for shipping was

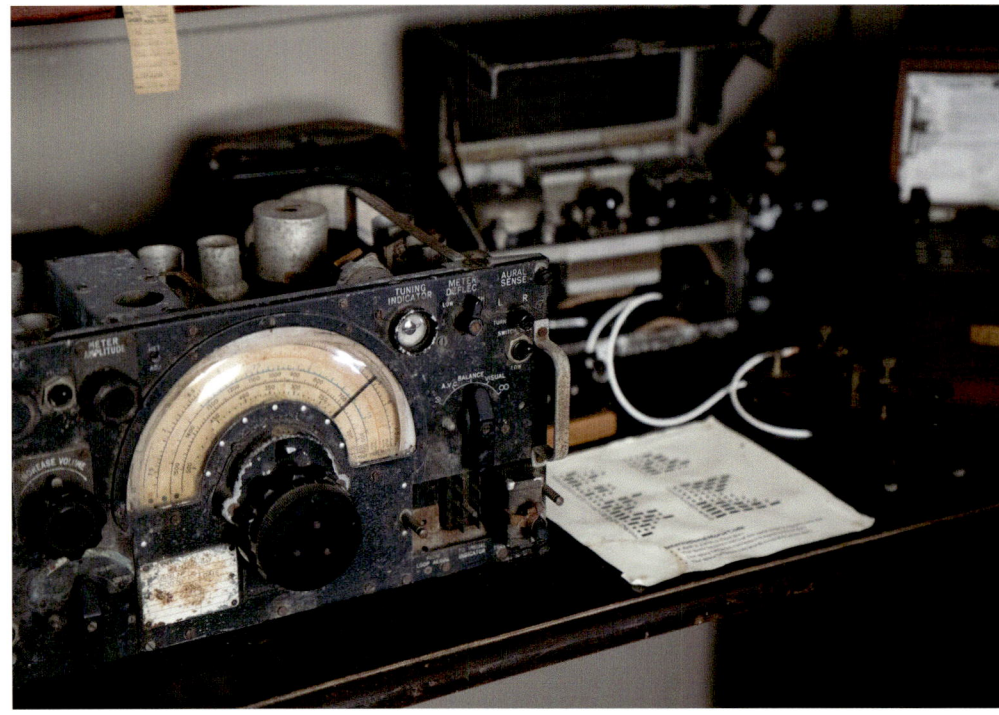

Receiver equipment at the Lizard Wireless Station (Image Copyright National Trust, with permission)

of the order of 50 miles (80 km), increasing to 100 miles (160 km) at night.

The Lizard Radio Station had two purposes. Its overt public purpose was to provide ship-to-shore communication. The secondary, more secretive, purpose was to test the viability of replacing subsea cables with powerful longwave transatlantic and transoceanic radio stations; this was the project that Fleming and Richard Vyvyan and George Kemp and about 100 local Cornishmen were busy working on in Poldhu, a few miles along the coast. The Lizard station was effectively a test bed to see what would happen to relatively low-power ship-to-shore radio in the presence of high-power transmissions from Poldhu.

Poldhu was a radio station built on an altogether different scale, not only when compared to the Lizard but to any other radio station anywhere in the world at the time.

The beating heart of Fleming's high-voltage empire was a 32-horsepower Hornsby-Ackroyd oil engine driving a 25 kW Mather and Platt alternator, producing 2,000 volts at 50 cycles per second. This was stepped up to 20,000 volts and fed to a closed oscillatory circuit in which a capacitor discharged across a spark gap via the primary of an RF transformer. The secondary of the transformer was connected to a second spark gap and capacitor, and the primary of a second RF transformer. The secondary of this transformer was in series with the antenna. Keying was achieved by short-circuiting chokes in the output of the alternator. The spark rate was somewhere between 5 and 12 sparks per second.

The capacitors were made up of 20 glass plates, each 16 inches squared, coated on one side with one square foot of tin foil, and immersed in stoneware boxes filled with linseed oil. The antenna system consisted of 400 wires suspended in an inverted cone from a 200-foot circle of masts, with 20 masts in all, each 200 feet high – all of which blew down in an autumn gale. Two revolving metal discs helped quench the spark. There were tens of tons of giant Leyden jar batteries.

All of this was put in place to send the letter S (three dots) across the Atlantic. After the gale, the ever-patient and beleaguered Kemp was instructed to build a simpler and more wind-resilient antenna structure, and he replaced what was left after the storm with two 150-foot masts

Workmen, Marconi Signal Station, Poldhu Cove, Mullion, c.1901 (from the collections at Kresen Kernow, Reference AD2548)

supporting 60 copper wires in a fan shape. The two towers were replaced in 1902 with four timber towers (see Appendix 5) then replaced in 1912 with steel tubular masts. The two mast antennas were installed in seven days – and, curiously, could have been the reason why the Atlantic Leap worked; two antennas with wires strung between them would have had some directional gain.

On the other side of the Atlantic, 22 masts, 100 feet tall, had been erected in Cape Cod, all of which, just as in Poldhu, blew down in a gale. Marconi decided to move the receiver to Newfoundland, which had the advantage of being 900 miles nearer to Cornwall than Cape Cod. The financial pressure on the company was extreme; £50,000 had been invested in Poldhu and Cape Cod, with no financial return. But the potential was enormous. The transatlantic subsea cable route was fabulously profitable, but a transatlantic cable cost $4 million, versus somewhere between $40,000 and $60,000 for a longwave high-power radio station. There was the small challenge of radio running costs, and the cost of oil, coal, peat or hydroelectricity, but the potential to eat the lunch of the subsea cable industry saddled with $400 million of (literally) sunk investment was more than interesting. However, cash was tight, and Marconi could not afford the time or money to build yet another antenna array. Instead, he found a usefully positioned hill on the coast of Newfoundland, called, appropriately, Signal Hill. An unprepossessing building provided a launch point for a kite or a balloon to suspend a temporary aerial.

He duly set sail from Liverpool with 2 tons of iron filings, several vats of sulphuric acid (for the hydrogen balloons) and some Levitor kites designed by Baden-Powell. An example hangs on the ceiling of the Marconi Centre in Poldhu.

There is substantial folklore surrounding the actual receipt of the message. It was daytime, so propagation was not ideal. It was windy; one balloon was lost, and the kites were wildly unmanageable. The whole exercise lasted an exhausting three days. Percy Paget was so tired and ill that he had to go to bed. On the afternoon of 12 December there were three pre-agreed transmissions from Poldhu at 12.30, 1.10 and 2.20.

At one point, Marconi heard three faint clicks (from the aerial connected to a kite), and handed the headphones to George Kemp, who agreed that he too could hear them. There were sceptics that said the setup was a sham, but on balance it is more than likely that it did happen. Marconi, who had brought a syntonic receiver with him, replaced it with an Italian coherer made by his friend Luigi Solari; this was a glass tube with a plug of iron at one end and carbon at the other with a blob of mercury in between, a forerunner of the semiconductor rectifier, 50 years later. As this was a broadband device, it could pick up anything from longwave to shortwave. Whatever harmonic energy was arriving at the wildly varying antenna would probably have been enough for the coherer to respond.[71] Edison was initially sceptical, but after talking through the whole procedure with Marconi, was convinced.

The Canadian government offered Marconi $60,000 to build a wireless station at Cape Breton: cable shares went down, and Marconi shares went up. Marconi, at 27 years old, hosted a banquet for 300 of the great and the good in New York, organised by the American Institute of Electrical Engineers. Alexander Graham Bell signed Marconi's menu card.

In July 1903 the Prince and Princess of Wales (later King George V and Queen Mary) visited the Poldhu Station, then travelled down to the Housel Bay Hotel for lunch, followed by a walk along the cliffs to the Lizard Wireless Station.

On Thursday 4 July 1912 the *West Briton and Cornwall Advertiser* reported on

> The international inspection of the Poldhu site by two hundred delegates from the International Radio Telegraphic Conference representing over 30 nationalities.
>
> The weather was delightful, and the visitors saw Cornwall at her best. Luncheon was served in a large marquee on the lawn of the Poldhu Hotel.
>
> Mr. Marconi and Mr. Isaacs (the relatively new Managing Director of the Marconi Company) guided the visitors over the station and witnessed messages being sent on behalf of delegates to Cape Race, Vigo, Madrid and Lisbon and messages of congratulation from the Acting Premier of Canada, the President of the United States, the Minister of Naval Affairs for Canada and the Acting Secretary of the American Navy. The delegates were brought from Paddington to Falmouth by special train and accommodated in half a dozen hotels. At the conclusion of their visit to Poldhu they returned to Falmouth, steamed up the Fal to Truro and were entrained for London at 4.00 pm.

Marconi and his team knew how to put on a show as and when it was politically and commercially useful.[72]

Poldhu continued as a ship-to-shore station up until 1923, when it became a research station under Charles Franklin (mainly working on shortwave beam radio, see above). The site was closed in 1933, and the site was cleared in 1935; six acres of clifftop were donated to the National Trust in 1937, with the remainder of the area added in 1960.

The granite monument was erected in November 1937 by the Marconi Company, and concrete foundations and earth structures remain as a reminder of what used to be there, including concrete anchor blocks for the mast stays, mast bases, the tiled floor of the transmitter building and a circular grass-grown 165-foot diameter rail track, which was where the Franklin beam aerial had been constructed in the 1920s.

The Marconi Centre was opened on the site on 12 December 2001, the centenary of the first transatlantic transmission, and it is manned every Sunday afternoon by a dedicated team of volunteers, including members of the Poldhu Amateur Radio Club. The museum includes the chair used by Fleming when he needed a rest from his high-energy labours. Any leisure time spent sitting in the chair was short lived. There were even bigger and more powerful high-power radio stations to design and build, with the mission to provide better radio links to the USA (from the West Coast of Ireland) and to the Empire (from Wales).

Long-distance longwave high-power radio on the west coast of Ireland (Clifden) and Waunfawr in Wales

The west coast of Ireland was less convenient than the Cornish coast, but a site was available in Connemara, 80 miles from Dromoland, the ancestral home of Marconi's wife, Beatrice O'Brien. The site had a lake for water (for the six steam engines) and a peat bog (for the steam engine boilers). This was going to be a radio station on an almost unbelievable scale, matching a station of similar size and power in Glace Bay in Newfoundland, known as Marconi Towers. A pattern of long wires stretched out above the ground, facing in the direction that the signals need to go in. The aerial system and huge earthing system were connected to a battery storage and condenser plant, the size of a four- to five-storey building, housing thousands of steel plates hanging from floor to ceiling. The power supply was a 15 kV DC generator (three 5 kV generators in series). Note the power source was DC. Standby batteries (6,000 2-volt secondary 30 AH batteries in series) at both stations were the largest batteries the world had ever seen. The eight wooden masts were each 210 feet tall and extended for half a kilometre.

At the heart of the Clifden/Marconi Towers Station was a whirling 5-foot spark discharge disc, with studs on its perimeter. Each time a stud passed between two electrodes, a 15 kV spark jumped the gap. The regular spark rate was 350 sparks per second. The power consumed by these stations was in the range of 100 to 300 kilowatts, and 150 kilowatts was fed to the aerials at 15,000 volts. The transmitters could be heard hissing and crackling several kilometres away.

In October 1907 transatlantic telegrams, now known as Marconigrams, started to be transmitted between the two stations at half the price of a cable telegram. Sarah Bernhardt sent a message. Three years later, in 1910, the Marconi Company paid its first dividend. But the Easter Rising in 1916 was to prove to be the beginning of the end for the Clifden site; Sinn Fein declared Irish independence in January 1919, and the Irish Free State was set up in 1922. Clifden was taken over by the IRA, and the site was used for public service broadcasting from 1 January 1926, using Marconi equipment. Appendix 4 provides additional information on the cable connections to and from the west coast of Ireland (Valentia Island to Heart's Content in Newfoundland) and the radio links between Clifden and Glace Bay.

In 1902 a young engineering graduate, Henry Joseph Round, joined the Marconi Company and worked at the purpose-built factory in Chelmsford. One of his early projects was to improve the efficiency of the high power radio site transmitters. He worked for Marconi up to the outbreak of the First World War and is credited with improving the transmission efficiency of the Clifden transmitters. He became a captain in the First World War, working in intelligence including radio direction finding (when Marconi was in Italy working on Naval Radio Systems). Captain Round is credited with alerting the Admiralty about the departure of the German fleet from Wilhelmshaven; this was the precursor to the Battle of Jutland on 30 May 1916. Some early tracking of Zeppelin airships was also achieved, as well as the development of simple but useful radio surveillance (Signals intelligence, aka SIGINT.) He re-joined the Marconi Company after the war, and worked on the Waunfawr project.

Having established a financially successful and stable transatlantic wireless telegram business, Marconi must have felt it was time to connect other parts of the world; Australia was high on the list of priorities. The Waunfawr transmitter was upgraded and coupled to a receive site a few miles away on the north coast of Wales. (Separating the low signal level receive path from the high power transmission signal increased the efficiency of the receiver.) The new transmitter, which went operational in 1918, had a 300-kilowatt single phase alternator coupled to two synchronous dischargers spinning at 3,000 rpm powered from the Cwm Dyli hydroelectric plant (built in 1906 to power the local slate industry) coupled to ten 400-foot masts.

This was the last hurrah of the longwave radio stations. Waunfawr continued to be upgraded through the 1920s with air-cooled and then water-cooled valves, and was retained as a back-up to the shortwave systems as and when they were affected by electrical storms. The site was also used to send pictures to the USA (from 1924) and was one of the assets transferred to Cable & Wireless in 1929.

Marconi claimed at the time that the Snowdon transmitter was the biggest in the world. Locals remarked that snow never settled near the site due to the heat from the power transmission lines – a snowless Snowdonia. By the mid-30s shortwave propagation issues on the transatlantic route were largely resolved, and in 1939 Waunfawr closed, eclipsed by the more efficient shortwave radio systems.

Shortwave makes big waves on Bodmin Moor

At the end of the war in 1918, Charles Franklin had been busy setting up a 12-kilowatt transmitter in Poldhu working at wavelengths between 30 and 5 metres (10 MHz to 60 MHz). Shortwave is defined as 3 MHz to 30 MHz (100–10-metre wavelength) so this was a shortwave system with VHF pretensions. On the other side of the Atlantic, Edwin Armstrong was working on the superhet and frequency modulation, but that is another chapter in another book.

In 1922 the *Nautical Gazette*, America's oldest shipping weekly journal, ran an article entitled 'Marconi's Utilization of Short Waves Means Great Advance in Wireless Art' (Volume 103, No. 1, 1 July 1922).

On 2 December 1924 the first Marconigram was sent from Poldhu to Cape Town by shortwave beam radio at 92-metre wavelength (3.3 MHz). This was the transmission that prompted the takeover of Marconi's long-distance radio assets, but more immediately was the moment when Marconi managed to persuade the UK government to adopt shortwave radio technology. The Bodmin Short Wave Beam Wireless Station was a crucial part of a new network of shortwave radio stations connecting Britain to what was left of its Imperial past.

The shortwave radio system had been tested in 1923 and 1924 between Poldhu and Marconi's yacht *Elettra* between the Cornish coast and New York. The sales pitch to the Post Office was a system optimised to

> establish communication with the Dominions using a fiftieth of the power, at a twentieth of the cost, providing a speed of working at least three times as great as that which was possible with the earlier long-wave system of communication.

The transmitters used paraffin-cooled power valves with copper to glass seals achieving an output of 11 kilowatts (compared to the 400 kilowatts of the Waunfawr longwave transmitter, or somewhere between 300 and 500 kilowatts at Clifden). The masts were 277 feet high, with cross arms at the top measuring 90 feet from end to end.

Bodmin Radio Station opened in October 1926 and provided links (beamed radio signals) to Canada, South Africa and Australia. The site was leased to the Air Ministry in 1940. After the Second World War the site was upgraded to provide communications with naval ships and naval bases overseas and in the 1980s it served for several years as the cabinet hotline to Moscow until the service was transferred to satellite Earth stations (including Goonhilly).

The remains of the station used to be visible just off the A30 in the grounds of a plant and tool hire company. Today the nearest large transmitter mast is at Carn Brea (Redruth) supporting digital broadcast TV, microwave backhaul and cellular radio.

The next 60 years

In the 1930s Marconi spent an increasing amount of time in Italy building a shortwave radio system for the Pope. The radio station, which opened in February 1931, was the world's first global broadcasting service, and the opening broadcast included a trumpet fanfare from the Vatican silver trumpets. There were seven broadcast transmitters in

Rome and 250 stations worldwide. The cost of the system was about $1 billion at today's dollar value. An updated version of the system is still operational. In 1933 a microwave link was established between Vatican City and Castel Gandolfo, 15 miles away. Marconi and Christina set off for a world tour, staying with the Roosevelts for seven days and meeting Mary Pickford, John Barrymore, Douglas Fairbanks and Charlie Chaplin.

Marconi died at 3.45 am on 20 July 1937 and was given a state funeral paid for by Mussolini; there was a mile-long procession with half a million people. At 6.00 pm Italian time, all 31 Cable & Wireless shortwave beam radio stations observed a two-minute silence.

In 1946 the Marconi Wireless Telegraph Company was taken over by English Electric. Then in 1963 the company was renamed as Marconi Company Limited. After the merger of English Electric with GEC in 1968, the company was rebranded GEC Marconi Electronics. In 1996 George Simpson took over from Arnold Weinstock, inheriting a cash mountain of £2 billion. But Simpson, an ex-British Aerospace manager, disliked the defence business, which was the company's main activity, so in January 1999 the defence business was sold to British Aerospace for £7.7 billion. The business, rebranded once again as Marconi, bought the US tech companies Reltec and Fore Systems, both of which were overpriced. At the same time Graham Wallace was steering Cable & Wireless into financial oblivion; Simpson found that his company's share price had halved almost overnight. In 2006 the bulk of Marconi assets were sold to Ericsson for $1.2 billion.

Marconi would doubtless be turning in his grave as the company that he had founded just over a century before disappeared in a sea of debt, redundancy and shareholder ire.

But sitting in the café by the beach in Poldhu Cove sipping our morning cappuccino, we need to remind ourselves that Marconi and his loyal team of radio engineers, working a few yards up the hill, had profoundly changed the world of communication, leaving a legacy that remains relevant today. His company was also going to play a pivotal role in the satellite industry just inland on Goonhilly Downs, something he would have been very proud of.

7

THE SPACE, SATELLITE AND DATA CENTRE STORY

What is going on in Goonhilly?

This chapter talks about the technology, engineering, economics and politics of the space and satellite industry with the Goonhilly Earth Station on the Lizard Peninsula as a working example. We look at how and why the Goonhilly Earth Station was built, what it was built for, why it is on Goonhilly Down, what it is doing today, and what it may be doing in the future.

After five decades as a government-owned facility run by the Post Office and subsequently BT, Goonhilly is now owned by private investors, including Charles Hargreaves, a co-founder of the stock trading company Hargreaves Lansdown. As such, the company is following in the footsteps of the ETC and Cable & Wireless (though Cable & Wireless was privately owned before it was government-owned before it was privatised) and the Marconi Wireless Telegraph Company. Effectively Goonhilly is a trusted partner to the government, providing a range of telecommunication services including hosting third-party services that are of importance to defence and sovereign security alongside commercial satellite communications, near-space and deep-space communications and space science. Space defence, cyber defence and data hosting are increasingly important activities.

In the first six chapters we have traced the story of the subsea cable industry along the Telecoms Coast, and the parallel story of ship-to-shore radio and long-distance longwave and shortwave high-power radio communication.

Far from becoming part of an industrial past, the Telecoms Coast in Cornwall is, now being reinvented as a springboard for the next generation of multi-purpose space and satellite communication systems.

A distinction needs to be made between the space industry and the satellite industry. The space industry is a global business supporting the exploration of Near Space, defined as anywhere within 2 million miles of the Earth, and Deep Space, anywhere further away. At time of writing (2025), the two manmade objects furthest away from Earth (also the fastest-moving manmade objects to date) are Voyager 1 and Voyager 2, launched in August and September 1977. The spacecraft are still talking to large dishes (the Deep Space Network) on Earth via VHF radios powered by radio isotopes. When the radio equipment fails, the spacecraft will carry on heading for the Oort clouds, two light years away. The journey will take about 30,000 years.

Near space includes the Moon. On 25 May 1961 President Kennedy announced to Congress that he was committing America to get to the Moon before the Russians. The Apollo missions achieved that goal, though at a significant cost to the American taxpayer. A new initiative, known as Artemis, the modern-day equivalent to Apollo, is – in theory at least – a coalition of the willing between America and whichever countries and companies America has decided are partners in the greater cause of global capitalism, which in this case includes Goonhilly. Marconi had to spend a lot of time being nice to politicians, and anyone running an earth station business must do the same.

Getting to the Moon is big business, but the commercial satellite communications industry is even bigger. Participating companies such as Space X (Starlink) and Amazon have enterprise values that are counted in trillions of dollars. Space X now has thousands of satellites in Low Earth Orbit. Blue Origin, a company largely owned by Jeff Bezos, the founder of Amazon, is launching a competitive network, Eutelsat is investing in upgrades of the One Web constellation and Chinese government agencies, working through proxy private enterprise corporations, are launching at least two LEO space networks each with thousands of satellites.

Compared to the 5 million or so cellular base station sites deployed round the world, these numbers are small – but huge when compared to a satellite industry that up to a few years ago counted the number of active satellites in space in hundreds, not tens of thousands. In terms of getting to space, at the time of writing, just one company, Space X, accounts for 80 per cent of global rocket launch capacity, China 10 per cent and the rest of the world the other 10 per cent. China is working hard to change this, and Europe is trying as well. Russia, however, still has the biggest rocket launch sites in the world, including the Baikonur Cosmodrome, leased from Kazakhstan. Cornwall has Newquay as a spaceport, which is modest by comparison, but easier to get to – and the locals are generally friendly.[73]

Low Earth Orbits: the new big market

Low Earth Orbit (LEO) constellations are typically deployed into orbits (described as orbital shells) between 300 and 1,200 kilometres. The orbits are low inclination (potentially anything from 0 degrees going round the Equator) to mid-inclination to high inclination (near-polar and polar, or over 90 degrees, including satellites that are sun synchronous or dawn-and-dusk synchronous, used for Earth observation). Most of these satellites go round in the same direction as the Earth (counter-rotating); these are known as prograde orbits. Some go in the other direction; retrograde orbits.

The antennas needed to talk to these satellites

From the perspective of an earth station, LEO satellites can appear anywhere on the horizon and from any direction. They then scoot across the sky and disappear over the opposite horizon. So instead of staring at a handful of geostationary satellites hovering 22,000 miles over the Equator, earth stations now need to track and talk to dozens of satellites travelling at 17,000 miles an hour (and occasionally faster). They could do this with the existing dishes – but this is not what the dishes were designed to do, and even if the motor systems were upgraded the maintenance costs would be problematic.

The alternative is to use flat panel phased arrays the size of a picnic table, with typically 1,024 antenna elements (or a similar large number suited to designing a rectangular array). The antenna elements are controlled by a signal processor to steer the antenna transmit and receive beams in order to follow the satellites as they move overhead. The steering is done by changing the phase of each antenna element, usually achieved by adding a delay to the signal. This is the equivalent of moving the antenna elements further apart, but is achieved by changing the electrical rather than physical distance. The antennas are normally housed in radomes, and in a typical LEO earth station antenna farm there can be several dozen of them. The same antennas are sold on an individual basis to subscribers; the antenna elements can be built into car and truck and tank roofs, and added to aeroplanes and ships and yachts.

This is not new technology, but existing technology that has had engineering attention. Hertz would be familiar with the radio principles, as would Round, Franklin and Marconi, although they would still be impressed by the size and mechanised sophistication of the factories (mainly in the US and China), which are producing thousands of these antenna arrays per day. It is, however, nice to know that the predecessors of these modern-day miracles of radio technology were being developed and tested 100 years ago on a windy cliff in Cornwall.

The dish is not dead and remains the best way to get a finely focused radio beam to and from space, including Deep Space and geostationary satellites.[74] The heritage dish antennas at Goonhilly still have a purpose, and if well maintained could last more or less forever. But they are now part of a broader mix of ground-based antenna technologies talking to satellites in many different orbits (defined by inclination and altitude).[75]

In 1904 Marconi dismissed Goonhilly as a possible follow-on site to Poldhu, on the basis that it was too boggy and wet. For the General Post Office (GPO), scouting for an earth station site in the late 1950s, however, it was ideal. Few people lived nearby, so radio noise was low, and as the site was a plateau with a flat horizon the satellite dishes had an unobstructed view down to a low elevation. It was also just over 100 metres above sea level.[76]

Husband and Company, a Sheffield-based company of consulting engineers who had worked on Jodrell Bank, were retained by the GPO to oversee construction. Local companies Peter Lind and Co. and E. Thomas Construction were contracted to help pour concrete.

Beginning with a beach ball

The launch of Sputnik in 1957 had caused a fuss, and America needed to play catch-up in the space race. A plan was hatched to launch a series of beach-ball-sized satellites into a highly elliptical orbit, which would mean the satellites would be above the North Atlantic for about 90 minutes before heading off into space again. These were the Telstar satellites, launched in 1962 and 1963, the years in which President Kennedy was wrestling with the Cuba crisis. At the Goonhilly site, Antenna Number 1, subsequently christened Arthur, was built. The BBC were brought in to oversee the VHF broadcasting transmitters and receivers, the GPO looked after the telephone circuits and a whole new generation of Marconi engineers worked to connect the whole system together. The Marconi Company, in various commercial incarnations, was to remain closely involved in Goonhilly for the next 30 years.

Transatlantic calls were made, pictures and TV signals were sent and presidents called prime ministers. Unfortunately, another team in the Pentagon were busy planning a high-altitude nuclear bomb test, Starfish Prime, which – along with other high-altitude tests (including Russian tests) – energised the Van Allen Belt, degrading Telstar's transistors,[77] a problem that LEO satellites still suffer from today.

What GES (Goonhilly Earth Station Limited) bought 50 years later was at first glance a bunch of old dishes in a bog. Arthur, the first and largest of the dishes, was (and is) Grade 2 listed. After Telstar, the big dishes at Goonhilly (at one stage there were six large dishes operationally active) were used to send pictures of Neil Armstrong stepping onto the Moon to a global audience of 600 million, the Muhammad Ali boxing bouts and the 1985 Live Aid Concerts, alongside phone calls, financial transactions over Intelsat and Eutelsat geostationary satellites and shipping distress calls (via Inmarsat). By the 1970s it was the largest earth station in the world, handling every UK international phone call.

Just like Porthcurno and Poldhu, it was an asset that was critical to national sovereign security.

Of the 60 heritage dishes, 25 are currently in use and, as we shall see, have a multipurpose future. Equally useful are the wayleaves and easements that were put in place back in the 1960s to get cables to and from the site. Establishing a right of way over and under the ground, even if you are a state monopoly, is a time-consuming and therefore expensive process but once they are there it is relatively easy to upgrade the cables running through the conduit or installed on old but still serviceable telephone poles. Goonhilly is therefore well connected to the subsea cables heading in and out of the Telecoms Coast (including the BT cables in and out of Kennack Sands), to GCHQ in Bude (covered in more detail in Chapter 8) and to the rest of the nation's fibre optic network.[78]

Goonhilly and the legacy of William John Bray

It is time to introduce William John Bray, also known as W.J. Bray or John Bray, as he preferred to be called (or Willy, as he was known to the young engineers working in the Post Office Engineering Department).[79]

Bray had much in common with William Preece, who had been the Chief Engineer of the Post Office in its Victorian heyday. Bray had been born in 1911, and – usefully for us – he wrote about his career and his life in his *Autobiography of a Communications Engineer*, Book Guild, London, 1999.[80]

At the age of 16 he built a spark radio transmitter and a crystal radio receiver. He joined the Post Office in 1935 and was appointed assistant engineer in the radio experimental branch in Dollis Hill in North London, where all the Post Office research work was carried out. He helped convert the Post Office transatlantic shortwave radio links (and other links) to single-sideband AM (reducing the amount of energy needed, increasing the bandwidth efficiency and capacity of the links and improving voice quality). The work included designing and building a steerable antenna. Karl Jansky was doing similar work on antennas for the Bell Laboratories in the USA, investigating radio noise; this was the start of the science of radio astronomy.

At the start of the Second World War a bunker was built at Dollis Hill (it was boom time for bunker builders). One of the projects at Dollis Hill (although not one that Bray worked on directly) was the balancing of telephone lines from the Chain Home and Chain Home Low radar stations, initially along the south coast of England, and then all around the British coast, including on Goonhilly Downs and a site near Newquay.[81]

After the war, Bray worked on the design and development of microwave radio systems and helped to develop the TV licence detector van. In 1954 he was appointed head of the Post Office's inland radio branch. A couple of years earlier a start had been made on a network of microwave relay towers, partly to help improve TV coverage but also to act as part of an integrated communications system for the Cold War. A chain of 14 towers, known as 'Backbone', was first mentioned publicly in a 1955 Defence White Paper: 'The Post Office plan is to build up a special network, both by cable and radio, designed to maintain long-distance communication in the event of an attack.' It was actually built in the early 1960s, in parallel with Goonhilly, as part of a larger microwave network bringing together civil and defence voice traffic, telegraphy, television and radar.

Almost every big hill along the Telecoms Coast has a microwave tower. The transmitter on the top of Carn Brea is an impressive example. These towers talked to another of John Bray's great works, the Post Office Tower in London, opened by Harold Wilson on 8 October 1965. At the time it was Britain's tallest building, 'a giant lighthouse above the London streets', with a total height of 620 feet, including a 40-foot mast, and it had what was claimed to be the fastest lift in Europe to take you to the famous revolving restaurant (planned to reopen soon as part of a new hotel development). Back then the Post Office Tower carried the BBC's new 625-line colour television broadcasts in and out of London alongside trunk telephone circuits.

All of which brings us back to Goonhilly. In 1961 Bray was appointed head of the newly created Space Communications System Branch. On 10 July 1962 Telstar, designed and built by AT&T and the Bell Telephone Laboratories, was launched from Cape Canaveral by NASA. A ground station had been built in Andover, Maine with a horizontal steerable horn aerial inside a 200-foot diameter radome.

There were five other ground stations (also known as earth stations) including Goonhilly and a station in Brittany, a Canadian ground station in Charleston, Nova Scotia, a German ground station in Bavaria, and an Italian station in Abruzzo. The spherical satellite, designed to fit into the hold of a Delta rocket, was 880 mm (34 inches) in length and weighed 77 kg (170 lb). While subsequent satellites had varying technical specifications, Telstar was spin-stabilised with solar panels generating 14 watts of solar power coupled to a travelling wave tube amplifier (TWTA) which sent a 4 GHz signal down to the earth stations (the relay of the uplink was at 6 GHz). The highly elliptical orbit was inclined at an angle of 45 degrees to the Equator with a perigee 952 kilometres (592 miles) from Earth and an apogee of 5,933 kilometres (3,687 miles). In each 2.5-hour orbit, transatlantic links could only be established for the 30 minutes when the satellite passed over the Atlantic. The first live TV transmission was received by Goonhilly from Maine at 3.00 pm on 23 July 1962, more or less 100 years and six months after Fleming had managed to send three dots across the Atlantic to Marconi and Kemp in Newfoundland. The Queen (Elizabeth, not Victoria) spoke of the 'thrill of communicating through a dot in space'.

The Post Office Engineering Department (based at 207 Old Street) designed and commissioned an 85-foot (25.9-metre) diameter steerable parabolic reflector weighing 800 tons, specified to withstand gale-force winds with minimal induced vibration. The dish was specified by Tom Husband, who had designed the 240-foot (73-metre) diameter radio astronomy aerial at Jodrell Bank (the dish that had successfully tracked the VHF signal from Sputnik in 1957). The new antenna had to track Telstar with a pointing error of less than 0.06 degrees as it moved across the sky at up to 1.5 degrees per second. Dollis Hill worked on the computer-controlled tracking system. Apart from getting the polarisation of the antenna the wrong way round (swiftly resolved), the TV pictures were of good quality. The popular TV broadcaster Raymond Baxter was on site at Goonhilly to record the occasion for the nation.

Writing in his memoir 38 years later (in 2000), Bray pointed out that by the turn of the century Goonhilly had become the busiest earth station in the world, with 15 large open-dish aerials named after the

Knights of the Round Table – who, as we all know, were based in Cornwall. The earth station also supported the Intelsat Early Bird satellites.[82]

In 1963 John Bray returned to Dollis Hill, and in 1966 was appointed Director of Research. He retired in 1975, having led the design and construction of the Post Office's new research centre at Martlesham Heath in Suffolk. Appendix 6 has more details on Arthur (Aerial Number 1) and Aerial Number 3 (Number 2 was controversially demolished). Aerial Number 3 was used as a prototype for the satellite dishes at Madley in Herefordshire, which is where the telephone circuits were moved in 2008.

The radio frequency future of Goonhilly[83]

As with the rest of the radio telecommunications industry, including cellular radio, as technology improves it becomes possible to move to higher radio frequencies. The advantage is more bandwidth (higher

Goonhilly today (Image Credit Goonhilly Earth Station Limited)

capacity) and more gain, from both parabolic dishes and flat panel arrays. It also becomes possible to have narrower beam widths. This reduces the amount of interference to and from other radio systems.

Markets also evolve; TV and telephone calls are now mainly delivered over subsea and terrestrial fibre. Geostationary satellites are, however, increasingly used to relay radio signals from LEO satellites back to earth stations, opening up new added-value opportunities for earth station operators.

The optical future of Goonhilly: from Land's End to the Lizard, from longwave to light

LEO satellites do more than communicate. Some of them study the stars; Hubble is one example.[84] Many of the others are Earth observation satellites; these take pictures of the Earth ranging from infra-red through visible (the red/green/blue that our eyes can see) through to ultra-violet. Small groups of satellites do RF trilateration, spotting where signals are coming from (useful for mapping troop movements and terrorists). Other satellites, arranged in a long line in space (typically 125 kilometres in LEO), image the Earth using a system known as Synthetic Aperture Radar (SAR).

Imaging can now be done from LEO at sub-metre resolution. The result is that exabytes of data[85] are now being collected in space. This data needs to come back to Earth, and there is not enough radio bandwidth available to do this. This is why optical free space links are now being used from earth stations to downlink data from space. The same links are also being used to route Internet traffic over LEO and geostationary satellites. Optical links work best at night in cool and dry conditions, and they work particularly well on high mountains. As with longwave and shortwave radio, it is always dark somewhere on Earth, so optical links can be dark path routed back to earth stations in, for example, Australia, Europe and Latin America. The optimum optical path to geostationary satellites (the most direct/shortest path) is from earth stations close to the Equator – but, as with radio signals, optical links to and from earth stations at higher latitudes (including Goonhilly) are likely to be needed. Most LEO communication satellites now have a mix of RF and optical transceivers, so by default earth stations are investing in integrated RF and optical ground station support infrastructure. As with solar panels and wind farms, optical free space links do not work all the time (hampered by snow, sleet, rain and fog, for example) but on a clear cool still night they can deliver throughput that is close to terrestrial fibre. They even work reasonably well during the day.

The (secure) data centre future at Goonhilly

All of that data from space needs to be stored somewhere and analysed somewhere, hence the need for a hyperscale data centre at Goonhilly. This is a Tier 4 facility, which means that is completely fault-tolerant, with redundancy for every component including backup power supplies. In other words, it is suitable for use as part of the United Kingdom's security-critical infrastructure, which in turn is part of a much bigger network that includes the US, European and antipodean (Australia and New Zealand) partners. As such, this is a continuation of a long-standing Goonhilly involvement in ultra-secure defence-related international communication systems starting in 1977, with Goonhilly very much in the loop of the early Internet, originally designed as a resilient packet-based communications routeing protocol for the Cold War.

In the early years of the 20th century, Porthcurno could claim to be the world's largest subsea cable station. Then, by the 1930s, Cable & Wireless owned and managed the world's largest shortwave beam radio system. Whether Goonhilly can scale the giddy heights of being the world's largest earth station going forward is open to debate; to compete with the earth stations being built in China and the USA (and Australia) it would need to cover most of the Lizard Peninsula rather than the 142 acres it currently occupies. This would probably be unpopular with the neighbours. However, as with Porthcurno and Poldhu (for a few years) and Bodmin, Goonhilly is politically important, and as flat panel arrays take up less space than dishes and everything gets smaller as frequencies increase, size is not as important as it used to be. Goonhilly is also well placed geographically and politically to take advantage of a whole new generation of space network machine learning technologies.[86]

Goonhilly, closely coupled to GCHQ in Bude, essentially becomes what Bodmin used to be, a hotline to the rest of the world, though with broader capabilities. This brings us to the last topic of this book and our final chapter, the Telecoms Coast security story.[87]

8

THE TELECOMS COAST SECURITY STORY

Tudor spies and the threat from Spain as a precursor to today's defence 'ecosystem'

In the 1560s John Dee (1527–1609), one of Queen Elizabeth I's spying team, signed off his secret letters to the Queen, using 00 ('for the Queen's Eyes Only') and 7, the lucky number for sorcery (Dee was also a sorcerer and magician). His day job was to find out when and where the King of Spain intended to land in England, with the Cornish coast as a convenient option. Signals intelligence in the Tudor age was largely a business of capturing and decoding encrypted written messages or – rather more effectively – getting spies into the Spanish Court or putting captured spies on the rack. Francis Walsingham (1532–1590), in his role as principal secretary to the Queen was the semi-official boss spy.

In principle, not much has changed since then. Although it seems unlikely that we will be invaded any time soon by an Armada of Spanish galleons and the rack has fallen out of favour, the business of sovereign countries spying on each other continues, the general rule being that it's okay to spy, but embarrassing if you get caught in the act. National defence is closely coupled to sovereign security. Warships still dock in

Plymouth, and the Royal Naval Air Station (RNAS) at Culdrose, 4 miles from Poldhu, is home to one of the largest helicopter bases in Europe engaged in anti-submarine operations, safeguarding frigates at sea and, more recently, managing drone operations. It is a busy airfield, and it is getting busier. The security of communication systems is a critical part of any fighting force, and the RNAS is no exception.

Before the First World War the airbase had hosted airships which, as the Germans knew, made excellent observation platforms. They were then used by the Germans in the First World War; my dad was born prematurely in a Zeppelin raid over Sydenham, in South London, in 1917.[88] Zeppelins used naval codes. All three German naval codes were captured during the war by the French, who shared the codes with Britain, so although it was nice to have direction finding, we already knew where the Germans were going most of the time.

Commercial and sovereign security – terrestrial and space integration

It is not just countries that spy on each other. Commercial espionage is as important as it ever was. Neil Maskelyne and his 170-foot pine mast on the cliffs above Porthcurno was just a small step for a rather strange man and an equally small step for the ETC in its mission to find out what Marconi was doing on the other side of Mount's Bay.

Today the AI and machine learning algorithms roaming through the extended datasets at GCHQ in Bude (and Goonhilly) look for links between commercial and military intelligence; an increase, for example, in the number of cars parked at a factory designing and making drones in Russia is a sign of bad things to come. This information is now available in real time from space.

The US Space Defense Force was set up in the USA in 2019. In relative terms it still has a small budget ($30 billion out of a defence budget of $750 billion), but a swathe of countries, large and small (over 90 at the last count) now also have their own space defence programmes and space agencies. A surprisingly large number of countries have their own spaceports as well (BryceTech https://brycetech.com/ keeps an updated map of the location of global space ports on its website).[89]

The ability to launch spy satellites from sovereign air space is increasingly regarded as strategically important.

China is at the forefront of the race to get thousands of spy satellites, including 'commercial' satellites carrying defence payloads, into LEO. Most of the Starlink satellites have defence-related systems sitting next to the radio and optical communication transceivers, which goes a long way to explaining why earth stations like Goonhilly are now classed as high-security defence assets. Laser weapons in space introduce another potential threat – or opportunity, depending on which side you are on.

Wireless versus wireline security

The idea that wireless was easier to eavesdrop was an egregious myth propagated by cable operators. In the American Civil War (12 April 1861–26 May 1865), the Confederates tapped into the 15,000 miles of cable that the Yankees had deployed to spy on 'enemy movements' and to send confusing and misleading commands back down the line. The Confederates still lost the war. Charles Wheatstone and his chum, Charles Babbage, the man generally credited with inventing the computer, both loved to do code cracking – more fun (and more useful) than crosswords. So once the Yankees realised they had been hacked they started using encrypted codewords.

Cryptography was also used to secure financial transactions. Each subsequent conflict introduced yet more complex coding and decoding schemes. In the Second World War the scientists and engineers at Dollis Hill designed the valve-driven computer that helped decrypt the German Enigma machines, with significant help from Alan Turing – and from the German operators, who regularly signed off their sessions with '*Heil Hitler*'; repeated words and phrases, known as a crib, can be an effective back door into communication security systems.

The crypto wars: post quantum cryptography and quantum key distribution (QKD)

Today, research teams are busy working on what is called Post Quantum Security, which uses quantum entanglement to secure messages; as photons (more correctly described as quantum particles) can spin clockwise and anti-clockwise at the same time, computers are no longer constrained to binary notation. Entanglement is an additional but closely coupled phenomenon where an entangled photon (quantum particle) changes spin state simultaneously with its doppelganger photon at any distance. The speed of light does not seem to be part of this curious process (described by Einstein as 'spooky action at a distance'). Google, Meta, and all the web scale majors, are invested in this new essentially photonic rather than electronic security domain. Their business depends on it.

As explained in Chapter 5, these companies, most of which did not exist 20 years ago – and none of which, including Microsoft, existed 50 years ago – now own most of the world's subsea fibre, with similar amounts of money being spent on high capacity RF, optical satellites (both uplinks and downlinks), terrestrial and space-based optical data storage. This is because the distribution of quantum cryptographic codes (quantum key distribution) can only be efficiently realised over optical (photonic) channels. This is not just for defence applications, but is also essential for all financial transactions, including cryptocurrencies such as bitcoin. In the longer term unbreakable cryptographic codes will need to be calculated using optical computers in order to be sustainably power-economic.

Taking people out of the loop: GCHQ in Bude replaces easy listening

To date, ways have always been found to decrypt voice and data. In the Second World War there were a host of direction-finding and listening posts. The St Erth Station – 6.5 kilometres (4 miles) south-east of St Ives, 10 kilometres (6 miles) north-east of Penzance and a mile away from the A30 in Hayle, 'a tranquil village centred around a pretty Church with a small river running through it' (the village, not the church) – was an important listening post. This was signal intelligence based on local people scanning the airwaves wearing headphones with a pencil and paper and a codebook to hand. Michael Griffiths' book, *Listening to the Enemy: War comes to a Cornish village, St Erth Radio*

Station 1939–1964, tells this part of the story of the Telecoms Coast in detail.

Today the heavy lifting is being done from space, with the data being sent down from LEO and then distributed either over subsea and terrestrial fibre or over inter-satellite RF and optical satellite links (including GSO relays; see Chapter 7). Where possible the data is being analysed by computer. The ETC and Cable & Wireless spent the first 50 years of the 20th century replacing operators with machines. Over the past 50 years the spying industry, to an extent, has done the same.

On 6 September 2024 GCHQ marked the 50th anniversary of the satellite surveillance facility on Sharpnose Point in North Cornwall, 200 feet from the Atlantic Ocean and 400 feet above sea level. The site now hosts over 20 dishes.

John Ferris, in his authorised history of GCHQ, claims that the station at Bude in the 1970s had a better intercept capability than any other UK antenna farm, collecting more information than every other GCHQ station put together. The 2024 press release states that GCHQ Bude 'remains to this day a site that is highly valued not only by our own organisation, but by our partners across the Five Eyes intelligence community'.[90] Telecommunications surveillance, like tin mining (in which 1 tonne of tin creates 99 tonnes of road chippings), requires sorting and sifting.

We may think of AI and machine learning as something new, but even back in the mid-1970s universities such as Harvard were offering PhD level courses on AI design and implementation. The theory was understood – it had been developed by those 18th- and 19th-century clerics, the Reverend Thomas Bayes (1702–1761), the father of conditional probability, and George Boole (1815–1864), the OR, NOT and XOR man behind every Google search – but the practice was more problematic. Augustus De Morgan (1806–1871), a near-contemporary of George Boole, is also relevant to present-day searching and sifting.[91]

Bude was the first serious attempt to automate this sifting process, using at least the underlying principles if not the current practice of AI. That said, a lot of people today still sit at desks sifting through the Internet – but it is better than turnip-picking on a cold windy rainy day.

Turnip-picking and telecoms

While we are on the subject of turnips, it is time to head off from the Land's End Hotel and travel a mile up the A30 to Sennen where a road to the left heads down to Sennen Beach, one of Cornwall's popular surfing beaches. Three hundred yards further on, a road with no signpost appears to the right. The absence of a signpost on any tarmac road is generally a sign that there is something on the road that somebody doesn't want you to know about or notice.

Flanked on either side by turnip fields (or other miscellaneous root vegetables), this road to nowhere turns out to be the digital highway to everywhere. In 2007 GCHQ started a project known as 'mastering the Internet,' a joint project with Five Eyes and a billion-dollar budget.

The first bit of the Master Plan was to build interception points close to where the long cables for instance from India came ashore, having crossed over from America. Most sovereign nations have lawful intercept rules and regulations. These regulations permit intelligence agencies to splice fibre optic cables at the point where the subsea cables become terrestrial cables. The intercept technology is relatively simple. If you can modulate an optical waveband, you can equally easily demodulate it by passing the light path through a signal analyser tuned to the particular wavelength/frequency of interest. (See **Appendix 2** on fibre technology.) There will then be some time division multiplexing which can equally easily be demodulated and as if by magic, you now have access to whatever happens to be passing down that particular light path in that particular time slot.

Some but not all of the cables of interest come ashore at Sennen Cove, carry on up the beach and a hill and end up in five buildings which no one is supposed to notice. The first site on your left could possibly be a cow shed except that cows don't need a non-interruptible power supply or multiple broadband interconnections (evident by looking at the manhole cover in the middle of the road a few feet away from the shed).

A few hundred yards further on to the right is a Water Treatment plant (also classed as Critical Infrastructure) and what may be a Vodafone intercept point some way down a private track with an electrified gate with a DO NOT ENTER sign.

Five hundred yards on there is a brutalist concrete building to your left on a sharp corner to the right; it was originally designed to be underground but the Gneiss Granite proved too expensive to excavate. Several small satellite dishes are also on the site. This is a BT site and is manned and marked as Private Property but if you are an admirer of Cold War era architecture then there is nothing to stop you taking a picture from the road or adjacent footpath.

Five hundred yards further on you come to a T-junction where you turn left to head down to the Porthcurno Valley. To your left is Skewjack Hut and to your right is a large blue building with a curved roof. These buildings are what are known as the Skewjack Farm site. The building with the curved roof, designed by Poynton Bradbury Architects and built by E. Thomas (by then owned by Mowlem), won a Cornwall Architecture Award in 2002 which is probably not ideal if you didn't want people to know you were there.

It is however a fine building of which the people that designed and built it are rightly proud.

FLAG British transatlantic submarine cable terminal (Image credit Poynton Bradbury)

This is Poynton Bradbury's description of the project:

Poynton Bradbury Architects were appointed by FLAG (Fibre-optic Link Around the Globe) (now Global Cloud Xchange) to design and deliver the terminal building for the FLAG Europe Asia (FEA) project, one of the largest bandwidth communications stations in the world.

The building was designed and constructed as one of four terminals serving a £1.8 billion 28,000-kilometre submarine fibre optic communications cable link that connects the USA, United Kingdom, Africa, Japan, and many places in between.

The building houses the ends of two submarine fibre cables, one that crosses the Atlantic to New York nearly 6000km away and the other cable that crosses the channel to France in Brittany.

Built by Cornwall-based Mowlem Building, the project won the 2002 CPRE RIBA Cornwall Architecture Award, with the judges praising the way in which the large high-tech complex had been integrated into the AONB landscape.

In 2014 a disgruntled USA intelligence contractor, Edward Snowden, revealed that this handsome building was the GCHQ interception point for the Reliance Communications international link from India. John Pender, from down the road in Porthcurno, would have been impressed by the technology, and not altogether surprised that the British government wanted to know what was going on in India. The rule of *Omerta* (it's okay to spy, but don't get caught doing it) had, however, been broken, and Reliance was rightly annoyed and upset.

In September 2018 the European Court of Human Rights ruled that the UK's mass data interception and retention programmes were unlawful and incompatible with the conditions necessary for a democratic society. Although it is a hard balancing act, our UK intelligence agencies today are as transparent as they can be about what they do and why they do it.

The secret and not-so-secret cloud

And in practice, life has moved on and real-time analysis of fibre cable traffic is now far less likely to yield useful intelligence. It is now more important to know what is being stored in all the world's hyperscale data centres. In 2014 Snowden claimed that GCHQ had sufficient storage bandwidth to store the world's Internet for a month. What he must have meant, I imagine, was the bit of the world's Internet designated as being of interest to GCHQ.[92]

There is a parallel investment by the same group of companies in space-based connectivity (LEO+MEO+GSO). In a way this is similar to shortwave beam radio, complementing and competing with the cable telegraph in the 1930s. The sums being invested are significant; Ali Baba, the Chinese Internet provider, has a war chest of £52.4 billion to spend on cloud computing and AI over the next three years (as of 2025) and a similar amount is being earmarked for LEO constellation deployment with data centres in space also a present ambition.

Physical security, burying cables under the sea: sea ploughs and the cable wars

It makes obvious sense to have two complementary delivery paths, space and subsea/terrestrial fibre, and the Telecoms Coast is well placed to provide both. In times of political tension, the game of digging up each other's subsea cables continues, with Russia and China in the frame for some recent deliberate destruction. Digging cables into the sea floor using remote-controlled sea ploughs, towed by and controlled from cable ships, helps a bit. We probably do the same to Russian and Chinese cables, but we don't hear about it on the news. Nature also has a habit of destroying subsea cables; in 1929 an underwater earthquake disabled nine transatlantic cables and in 2024 Tonga was disconnected from the Internet for several weeks due to a subsea volcanic eruption.[93] It is, however, hard to destroy hundreds of subsea cables and potentially tens of thousands of satellites in LEO. This is the rule of large numbers applied to resilience.

Tin mining and telecoms

All of which brings us to the end of our story; it is time to go back to where we started, with the Geevor Mine investing at the beginning of the 20th century in a more energy-efficient plant for the underground work and the mill, where the tin is relentlessly shaken and washed out from thousands of tons of granite rock. A few miles down the coast at Porthcurno, banks of lead acid batteries are being carefully maintained to connect Cornwall to the rest of the world through tens of thousands of miles of subsea cable. Across the bay, Ambrose Fleming and Richard Vyvyan are coaxing a few more thousand volts from their glass capacitors, firing hundreds of thunderbolts a second into the aether.

This is a story of power measured in watts, volts and amps, but it is also the story of the political power that accrues from a combination of power-efficient telecommunications, energy security and locally available raw materials. It is a sobering thought that tin and copper prices peaked before each of the two world wars, and as I finish writing this book they are once again heading up again in value. The potential reopening of South Crofty is an indication that politicians are becoming more aware that we cannot always rely on external non-sovereign third parties to supply critical metals and minerals, especially when political tension is high. It is also advantageous to have end-to-end control of telecommunication systems, or at least the ones that are critical to sovereign security.

The one thing we can be sure of is that by virtue of a combination of geography, geology and oceanography, the Telecoms Coast in Cornwall will play just as important a role in the security of our sovereign future as it has in the past.

APPENDIX 1:
GEEVOR ENGINEERING AND ENERGY REQUIREMENTS

Geevor provides a unique case study of a 20th-century tin mine built on the site of mine workings that go back for many centuries.

The submarine mine shafts required power for dewatering and also required compressed air for ventilation and power for rock-drilling equipment and for moving miners to and from the surface. Nature breaks down exposed and underground granite through water erosion and weather erosion and in river valleys it deposits sand, which can be hand-sifted to recover metals of different weights using a sieve and mining pans, hence the term 'panning for gold'.

A mill does the same, but in a small fraction of the time. When Geevor Tin Mine was set up in 1911, it was a state-of-the-art mine with a mill that was at the cutting edge of rock-crushing and rock-sorting efficiency, including electromagnets to pull out any old iron, nickel or cobalt.

The yield rate was adequate rather than spectacular (1 tonne of tin for every 100 tonnes of rock), so sourcing power and using it efficiently was essential to the economics of the mine. Bear in mind that just down the coast and in Mullion across Mount's Bay at the same time, Ambrose Fleming, a master of the AC/DC world, was busy working on the transmission efficiency of longwave radio transmission and Charles Franklin was busy working on antenna efficiency, including shortwave beam radio.

When the mine closed, some of the mill machinery was sold off, but most of the equipment is still in place, including the rock crushers, sorters and sifting machinery. Also in place are the cable and switch circuits handling the power, initially from the Cornwall Electric Power Company of Hayle, and then, from 1933, the 50-cycle supply from the National Grid. In the early 1950s a new 33-kilovolt to 11-kilovolt substation was built at the edge of the site, connected by underground cable (11-kilovolt three phase supply) to a switch house, where the power was converted to 440 volt three-phase for distribution to all parts of the mine. The power consumption was of the order of 2.2 megawatts per day, supporting a three-shift system, with the overnight shift doing mainly maintenance and lifting of ore to the surface.

During the 20th century this power supply, combined with many thousands of hours of hard and dangerous manual labour below and above ground, produced 50,000 tons of black tin and a profit of £7 million. On average over a million gallons of water, a quarter of which was salt water, was pumped from the mine every day.

What happened to the other 99 tons of (mainly granite) chips?

After everything metallic and semi-metallic (arsenic) had been sifted, crushed, sieved, washed and/or heated out of the stone chips, they were sold for roadmaking or concrete. Granite blocks were generally mined from surface quarries (Newlyn is referenced in Chapter 2). Cornish granite can be found in roads, viaducts and bridges. London's Kew Bridge was built in 1890 out of Cornish granite.

APPENDIX 2:
OPTICAL FIBRE TECHNOLOGY

Today most, but not all, long-distance (terrestrial and subsea) fibre is single-mode. In single-mode fibre the light wave propagates parallel to the axis of the fibre; this means there is no modal dispersion (more about this in a bit) with reach limited by polarisation loss and chromatic dispersion. The big leap forward in fibre capacity has been the shift from on/off keying (OOK) via two- or four-level intensity modulation to coherent modulation, in which information is phase-modulated onto the optical carrier. This transition has been enabled by the ever-improving performance of digital signal processors at either end of the cable and is similar to the gain achieved in copper cables using ADSL or VDSL. The default frequency is optical C band (191–195 THz), though super C band is now being deployed, adding 3.5 THz of pass band. Channel spacing can either be 100, 50 or 25 GHz. A few fibres packed together means that the cable is easy to handle, the outer cable remains flexible and the bending radius remains relatively tight, whereas more fibres mean less flexible cabling and a more complex termination process.

Single-mode and multi-mode fibre both work on the principle of total internal reflection. A core of doped glass is surrounded by an outer core of clear glass, the glass having a higher refractive index than the core. In multi-mode fibre, multiple beams of light at the same wavelength (but with different angles of reflection), are reflected back from the boundaries of the core. This means the beams have different path lengths and are thus separated in time and can be demodulated as separate light paths.

As with single-mode fibre, individual multi-mode fibre can be bundled together into an outer cable. Up to 6,912 cables can be packed into an outer diameter of 1.14 inches, though the cables need to be ribbon-spliced, which introduces losses. Optimised single-mode fibre (single-beam direct path) can have losses below 0.2 dB/km. In 1970 typical cable loss was 20 dB/km, which goes a long way towards explaining why and how fibre has become increasingly cost-effective over time.

APPENDIX 3:
CABLES AND BATTERIES, LINK BUDGETS AND NETWORK OPTIMISATION

Porthcurno Engineering training documents

The technical documents library in the reading room at Porthcurno includes training documents for engineers who were being sent out to almost anywhere in the world and needed to understand how to look after potentially dangerous equipment – which particularly means lead acid batteries.

Subsea cables in the Victorian era through the Edwardian era, and legacy cables for many years after that, were low-voltage (due to the low breakdown voltage of the cables) so were relatively safe compared to the high-voltage systems driving Ambrose Fleming's spark gap transformers or the 11-kilovolt three-phase supply systems powering the Geevor Tin Mine or the travelling wave tube amplifiers (TWTA) feeding RF power to the dish antennas at Goonhilly. However, the batteries used in subsea cable systems were, to mix metaphors, a health and safety minefield.

Banks of lead cell batteries were either trickle-charged (known as floating batteries) or put through a daily charge cycle. New battery cells also needed careful commissioning.

As and when stations had moved on to rechargeable batteries (dynamos for DC and alternators for AC from the 1870s onwards), they would be recharged either using a hot bulb engine (Edwardian era), internal combustion engine (ICE) or a diesel-powered generator. On-site power was a mix of 120 volt nominal DC power for functions such as the switchboard, 240 volt single phase and 400 volt three-phase. Internal combustion engines and diesel engines were used for site back-up power and a non-interrupt flywheel managed short-term power supply interruptions (allowing time for diesel generators or petrol engines to be started and warmed up).

Porthcurno had two Ruston diesel generators. The 1941 generator supplying power to the bunker is still in its original position.

Batteries were originally placed in the main building in Porthcurno next to the billiard room but in 1925 the batteries were moved to a fireproof accumulator room on the other side of the road. Porthcurno was in effect being used as an example of good practice, which could be replicated around the world. Given that a charge cycle was completed when gassing had been observed for two hours – the point at which energy from the accumulator is separating hydrogen and oxygen from the electrolyte (distilled water) – it was not a good idea to have people smoking in the billiard room, next to the batteries.

The lead acid batteries were usually open-top cells with glass rods as separators. With use, the positive plates oxidised to lead peroxide (a chocolate-brown colour) and the negative plates reduced to spongy lead (slate grey).

Recharging through the electrolyte caused ions to move in the reverse direction (the same principles apply to modern 12-volt lead acid batteries used in cars).

The low limit of the discharge was 1.8 volts per cell. A deep discharge could cause a heavy coating of lead sulphate. Full charge was indicated by a voltage of 2.5 to 2.6 volts per cell.

Specific gravity decreases 0.004 per cent with every 10°F. Calculation of the required ratio between the sulphuric acid and water was done at 60°F. The calculation would need to be adjusted for hot and cold climates. The use of distilled water was essential.

There were obviously problems handling sulphuric acid and a need to understand the explosion risk inherent in gassing. These are the instructions:

> When mixing electrolyte **ALWAYS pour acid into water using an earthenware or glass pot, stirring at the same time.** The Voltage of the mains is to be at least 2.7 times the number

of cells. Charging rate is 3 amps per square foot of possible plate surface.

The stand by battery in stations where AC is available should be continuously charged on a trickle charger with 100 mA charging current for a 100 Ah battery. **Severe overcharging may cause the lead plates to buckle.** Electrolyte should be kept up to half an inch above the top of the plates by addition of distilled water to make up for evaporation. The terminals should be coated with Vaseline.

NEVER INSPECT ACCUMULATORS WITH A NAKED LIGHT. HYDROGEN AND AIR ARE EXPLOSIVE. The accumulator room should be large, airy and cool with no direct sunlight. Commence the first charge immediately after putting in the acid. Charge should be at half normal rate for 12 hours then full rate. First charge should last 40 hours continuously. **Charging is not complete until vigorous gassing is observed for two hours.**

The document is from the 1950s but is undated. It can be read in full in the Reading Room in Porthcurno.

Footnote: Due to the low breakdown voltage of the early cables, throughput gain had to be achieved through improvements in receiver sensitivity, the mirror galvanometer being an early example. Receiver performance is equally important to all the other radio and cable systems covered in this book. Examples include Marconi's Maggie Receiver, low-noise low-loss front ends in the satellite dishes at Goonhilly (and satellites in space), low-noise low-loss front ends in 4G and 5G cellular radio equipment and smartphones, and optimised fibre optical receivers including free space optical front end lenses and filters for earth-to-space, space-to-earth and satellite-to-satellite links.

Link budgets

Three essays in the RSA book *Semaphore to Short Waves* provide insight into how subsea cable performance improved between 1870 and the end of the Second World War, and the basis on which parallel improvements in radio performance were realised from 1900 onwards.

Faraday, Maxwell and Field Theory by Frank A.J.L. James documents how the understanding of wave form theory evolved in the 19th century and how this informed engineering improvements.

For radio engineering, James Clerk Maxwell (1831–1879) provided the theoretical base on which Heinrich Hertz (1857–1894) and Oliver Lodge (1851–1940) developed the basic building blocks of radio communication.

Maxwell 'mathematised' the concept of the electromagnetic field developed by Michael Faraday (1791–1867).

Insulation for an Empire by Bruce J. Hunt shows how an understanding of materials testing, including gutta percha and copper, improved performance and performance predictability.

Latimer Clark (1822–1898), Charles Bright (1832–1888) and Fleeming Jenkin (1833–1885) and their work for the British Association Committee on Electrical Standards in 1861 established the basis for electrical measurement, including resistance (ohms), current (amps), potential difference (volts) and capacitance (farads and microfarads).

Jenkin's measurement of the electrical performance characteristics of gutta percha and copper in various thicknesses and various manufacturing methods allowed cable engineers to calculate what fraction of the total current would reach the far end of the cable, and what fraction would leak out laterally through the gutta percha. Signal rate though the cable was established as the direct product of the resistance and capacitance of the cable. As an aside, copper cable leakage allows cable traffic to be intercepted and decoded by submarines, whereas optical cable must be spliced to be intercepted.

Gateway to Empire by J.E. Packer explains how engineering improvements through the 20th century, including automatic transmission and regeneration, improved throughput, increased consistency and accuracy (word error rate) and reduced cost.

Outgoing signals would be at anything between 30- and 80-volts amplitude, with incoming signals of the order of a millivolt. From the 1890s duplexing required cables to be balanced against an artificial line, a long cable in a room full of heavy capacitance resistance boxes. Cables needed to be rebalanced at least daily, and sometimes more frequently.

Attenuation and distortion obeyed a square law; doubling the distance of the cable meant a fourfold reduction in signalling speed. With the introduction of regeneration from 1923 onwards, each incoming signal was sampled and re-signalled at more or less the original shape and amplitude. The regenerators were slaved to a master pendulum clock and a 30 Hz timing reference with sampling done at the point of least distortion. Between 1923 and 1927 regeneration was introduced at 120 stations, including Porthcurno, spanning 145,000 miles of cable. At a typical relay station, 30 operators working shifts would be replaced by six engineers with oil cans. John Packer makes the point that regeneration was an early example of network synchronisation, jitter control, bit interleaving and store and forward.

A reading of these sources is recommended.

Network learning

As more subsea cables were laid and as subsea cables increased in length, more information could be gathered and analysed and used to improve cable specification and cable network performance. Modern examples of this technique can be found in 4G and 5G networks and LEO satellites, where performance is analysed using machine learning algorithms, a discipline sometimes described as Network Learning.

APPENDIX 4:
VALENTIA ISLAND CABLE STATION

The other Telecoms Coast

In 1866, four years before the cable to Bombay was connected to Cornwall via Porthcurno, communications were established between Valentia Bay in County Kerry on the west coast of Ireland and Heart's Content in Newfoundland, a distance of just over 1,600 nautical miles (3,000 kilometres).

Although John Pender was to become an investor in the cable, it was initially financed by the American millionaire Cyrus Field, and could either be regarded as a catalogue of early disasters (four failed attempts to set the cable) or a steep learning curve, which William Thomson (later Lord Kelvin) used to further the understanding of subsea cable propagation and more or less everything else that formed the basis of a successful subsea telegraph industry.

The Valentia Cable Station remained operational from 1866 to 1966. Like Porthcurno (1870–1970), it therefore provides a 100-year technical and commercial case study of the subsea cable industry. Arguably it could have been more important than Porthcurno, spanning a shorter distance across the Atlantic – but, as we shall see in this Appendix, Irish politics intervened.

Valentia Island is a potential UNESCO World Heritage site. Updates on the status of this process are available on the Valentia Cable website, www.valentiacable.com

There is also an excellent book by Donard de Cogan, *They Talk Along the Deep, A Global History of the Valentia Island Telegraph*

Cables, published by the Radio Society of Great Britain in 2023, which tells the story of the station in detail.

The map on the Valentia Cable website heralds the cable as the Eighth Wonder of the World; 100 nautical miles north and almost 50 years later in 1906, Marconi was busy building what I am sure he would have been happy to call his Ninth Wonder of the World, spanning the Atlantic from the Clifden steam-powered longwave radio station.

The Valentia adventure started with Frederic Newton Gisborne (1824–1892), a British inventor and electrician. Gisborne settled in Canada in 1845 and was appointed Superintendent of Telegraph Lines in Nova Scotia. In 1852 he laid a deep sea cable (the first deep sea cable in North American waters), between Prince Edward Island and New Brunswick.

In 1854 he met Cyrus W. Field in New York. Unimpressed with terrestrial and local subsea cable investment opportunities, Field became interested – one might almost say obsessed – by the idea of a transatlantic cable. The proposition was simple. It took 14 days to send a message by boat from Liverpool to New York., whereas a message sent by cable would be more or less instantaneous.

In 1855 Matthew Fontaine Maury (1806–1873), a naval officer, had published the first comprehensive work on oceanography, *The Physical Geography of the Sea*. This work drew on measurements made by a naval officer, Lieutenant-Commander Otway Berryman from the USS *Dolphin*, which suggested the existence of a telegraphic plateau between Newfoundland and the west coast of Ireland, a myth that was to persist for the best part of 70 years.

Convinced that the Gods of Oceanography were set fair for the project, Cyrus Field came to London in 1856 and met Charles Tilston Bright and Isambard Kingdom Brunel. Brunel took Field down to Millwall to show him the SS *Great Eastern* under construction, prophetically pointing out how suitable the ship would be for the long-distance laying of subsea cable.

The seabed turned out to be bumpier and hillier than expected. The silver lining in this particular cloud was that the paying-out mechanisms on cable ships underwent a process of intensive improvement, as did the mechanical design of the cables.[94]

The early electrical specification of the cables was equally fraught. Dr Edward Orange Wildman Whitehouse was an English surgeon by profession, appointed Chief Electrician on the cable project, much to the frustration of William Thomson (retained at that point in an advisory role). Whitehouse managed to make sure that even when the first cables were laid; they either didn't work or rapidly failed, incurring additional costs of at least £100,000.

William Thomson, who was born in Belfast on June 1824 and died in Ayrshire 1907, had a clever (and dominating) father, who lectured first in Belfast then at Glasgow University, and who had studied Laplace and Fourier.

William, age 10, and his brother James, aged 11, matriculated at the University of Glasgow in 1834. By the age of 17, William was working to translate Fourier's mathematical analysis of heat flow to electricity flowing through a wire. The Chair of Natural Philosophy (later known as physics) became vacant in 1846, and at age 22 William took up the role, staying there for 53 years until he retired in 1899.

He was responsible for some of the most important technical improvements in subsea telecom cable technology and for the step change in understanding subsea signal propagation. He also risked his life several times during the laying of the first transatlantic cable.

The theory included the use of Fourier's maths as the basis for understanding pulse propagation, with the retardation of signals dependent on the resistance and capacitance of the cable. This was in accordance with the law of squares in which resistance is inversely proportional to the cross-sectional area of the copper core and dependent on the purity of the copper, with capacitance dependent on the ratios of the radii of the inner and outer conductors: the larger the inner diameter, the lower the capacitance, the lower the capacitance, the lower the signal retardation.

The practice included Thomson's patented telegraph receiver, the mirror galvanometer in 1858, for use on the Atlantic cable. The siphon recorder, sibling of the galvanometer, came along in 1870. Thomson also worked on how to reduce the impact of earth currents on these devices, which came from an understanding of what was actually happening. With a single-wire system, not only would the earth at either end of the cable usually be at a different potential, but the potential difference would vary during periods of intense solar activity and thunderstorms, moving the light spot on the mirror galvanometer. Compensating with resistors reduced receiver sensitivity (by introducing insertion loss).

In two-wire systems, connecting both cables together to form a loop eliminated earth currents. These systems could achieve 90 words per minute and were generally more efficient than the other option of adding a capacitor (condenser) between the galvanometer and earth.

Essentially, Thomson developed the principles of power-matching on the transmit path and noise-matching on the receive path – the forerunner of the Smith Chart introduced by Phillip Hagar Smith in 1938 and now the basis (complex impedance matching) of more or less all telecommunication transceiver front-end design.

Thomson also substantially furthered the understanding of how to match subsea cable to terrestrial cable and adjust the matching over time. Earthing of the submarine cable on the transatlantic cable was achieved by setting 2 miles of cable out to sea terminated by a large mass of zinc. Landlines were earthed by a rod hammered into the ground, with efficiency reducing in long hot dry summers.

Irish politics and issues of sovereign security

In 1869 the Valentia Cable Station employed 14 cable staff and 5 land staff. By 1900 the number of operators had increased to 40, then 200 in the First World War. But after that, numbers reduced, partly due to mechanisation but mainly due to politics.

From the earliest days of the station in 1866, there had been local outbreaks of violence associated with Fenian groups working towards Home Rule and the settlement of the Land Question.

Cable traffic was regarded as being security sensitive. In the First World War cable intercepts had enabled the British government to make the US government aware of German plans to support a Mexican invasion into the USA north of the Rio Grande, and a potential incursion by the Japanese into California. Together with the sinking of the *Lusitania* in 1915, this brought America into the war (in 1917). The cutting of German cables at the beginning of the war had been relatively easy, as British ships owned by British companies had set the cables and therefore knew where they were.

The British government began to realise that a significant number of operators at the Valentia Cable Station were likely to be Sinn Fein supporters. This became of particular concern after the Easter Rising in 1916.

In the Irish Civil War in 1922, land telegraph cables were cut and the station batteries were destroyed by the IRA. Direct subsea connections had been established from Valentia Island to Penzance in 1922, but these were vulnerable to sabotage. In 1923 a request from Western Union (the US-owned cable operator) to land a new cable from Sennen Cove to Valentia Island was refused by the Irish government.

Between the wars improvements were made by Western Union, including the introduction of high-speed loaded cables (inductance coils along the cable) and the use of Permalloy (high permeability nickel iron) and Mumetal (copper nickel iron alloy). Successors to Western Union were still paying landing fees to the Irish government up to 1982.

Other Telecom coasts

To an extent all coasts are telecoms coasts, supporting a mix of cable, radio and radar assets. Notable defence-related heritage radio sites can be found on the east coast of Britain – for example, Cobra Mist at Orford Ness and Fylingdales, both designed to provide early warning of an attack from the east.

APPENDIX 5:
MARCONI ARCHIVE, BODLEIAN LIBRARY: POLDHU

The Marconi archive at the Bodleian includes letters to and from Marconi and his management team (including Jameson Davis, Major Flood Page and Cuthbert Hall). Some of these are terse and necessarily formal. The archive also includes letters to and from the engineering team (Ambrose Fleming, George Kemp, Percy Paget and Richard Vyvyan). These are altogether less formal and show that there was a high level of mutual respect and forbearance in the face of many practical setbacks. The letters essentially show why Marconi managed to make things happen faster than Fessenden, Lee de Forrest and Professor Slaby.

A graphologist would most likely describe Marconi's letters as written by a man in a hurry, barely decipherable but to the point and almost always positive, even when grappling with either an existing or an imminent setback.

The letters include correspondence from Major Flood Page regarding the various possible sites along the coast, their relative advantages and disadvantages and their rental or purchase costs and access. The headland above Mullion Cove was preferred, but Poldhu proved easier to negotiate.

The archive is extensive and for brevity I have summarised the correspondence directly addressing the Atlantic Leap and the role of the Poldhu Radio Station as a development template for other high-power radio sites (including Glace Bay and later Clifden, Waunfawr and Bodmin).

The relevant manuscript (MS) references at the Bodleian are MS Marconi 187, 190, 197, 232, 249 (Maritime Marconi), 807 and 808 (Charles Franklin beam radio development in Poldhu).

The early letters are addressed from Fleming when he was at the Pender Electrical Engineering Laboratory in Gower Street, founded on a bequest from John Pender. (The laboratory was part of University College). The discussions include how equipment can be sourced at a competitive price including second-hand if available. Ambrose Fleming becomes exasperated by the unreliability of some of the equipment, including the oil engine; this probably at least partly explains the decision to use steam power at Clifden.

> April 26, 1901 University College, London, Remarks on the Marconi Station at Poldhu
>
> I have been at Poldhu for the last three weeks and have got all the plant in working order and have obtained signals from it. For the guidance of Mr Vyvyan in arranging the American Station (at that point being built in Cape Cod) I send the following remarks.
>
> We have had great difficulty with the Hornsby Oil Engine in getting it to start; four times out of five it will start all right, and the fifth time it will take a couple of hours to set it going …
>
> I have even gone as far as to consider the possibility of adopting a Steam Engine at Poldhu as I find that water could be obtained from a reservoir some way off. Mr Vyvyan should examine the possibilities of the site in America with a view to getting water, in case it should turn out that it is necessary to adopt a steam engine.
>
> Another of the difficulties that has presented itself at Poldhu is the great drop in voltage which takes place on the alternator at the moment of making the signal. The Mather and Platt alternator at Poldhu has of course a large inductive drop and there are no compensating coils in it, as in the case of the Westinghouse alternator. At the moment of making the signal, we take out of the mains a large wattless current which charges

up the condensers, and although this does not make a call on the engine for a corresponding amount of power, it causes a large drop in the alternator volts, and this is the difficulty that I am dealing with specially at the present time. …

On the other hand, the inductive drop in the Mather and Platt machine makes it very safe to use. In the course of my experiments, I have twice short circuited the machine, once when a condenser went, without blowing the switchboard fuses.

Note that there was substantial management concern over the cost and risk of the Atlantic Leap Project, including the opportunity cost. Briefing letters to the board from Major Flood Page acknowledge that other business opportunities, including potential Admiralty contracts, were starved of funding to keep the Atlantic Leap project on course. The loss of the aerial arrays at Poldhu and Cape Cod in stormy weather caused dismay. George Kemp, cited (by Cuthbert Hall) as being responsible, was defended as the best man to build things, but not necessarily the best man to design things. These early letters also document the ongoing debate about wooden masts (many people, including George Kemp, had substantial experience in building large wooden structures), or hybrid metal and wood – and then, four or five years on, a transition to metal masts once everyone was confident that they could be designed to be electrically safe. Metal masts had the significant advantages of offering a smaller cross-section to the wind and more consistent material quality, the sourcing of logs being generally problematic.

By December 1901 the specification of the new four-mast system at Poldhu was well advanced. Wood was still the preferred option:

> The proposed towers are timber deals, tied together by plates of wrought iron, bolted in position by suitable bolts so as to form a Skeleton Tower, 215 feet in height, three feet six inches of which is buried in concrete foundation … the four towers are placed in a square, the sides of which measure 210 feet, centre to centre and are designed to carry an eight part aerial, each part of 60 lengths of 7/22 copper wire attached at top and bottom to one and a half inch steel wire triatics …

APPENDIX 5

Wooden aerial towers under construction at Poldhu, 1902

Poldhu four-tower design wooden aerials under construction 1902, photo taken from the field edge, with the Poldhu Hotel in the background. (With thanks to Oxford Bodleian Library MS Marconi 1782 GEC Marconi 65114)

A foot of concrete is composed of 3 parts of beach shingle, 3 parts broken blue stone and 1 part Hollicks Portland Cement … the whole surface inside is covered with a concrete floor about 9 inches thick sloping from centre at an incline of 1 in 14, in order to drain all moisture from the foot of the towers.

The structure or structures are to be built on a cliff facing West on the coast of Cornwall, near Mullion at a spot known as Poldhu, the cliff being approximately 100 feet above sea level and the sea washes its base. A clear way is open to the Atlantic to the South West and the full force of the wind is felt. The prevailing winds are SW, W and NW. The strongest measured is believed to be 92 miles per hour.

Within the four towers a huddle of buildings in the middle housed an engine house, an oil store, a battery shed, a condenser house, a rigging shed, a jamming station, a carpenter's shop, a transformer house, residential quarters, a boiler house, a pumping station, a water reservoir, a blacksmith's shop, a cook house, a motor house, a coal house and general stores, a wireless telegraph office, a landline telegraph office, an engineer's office, a fitting shop and a paint shop.

Debates were ongoing about how to design and build earthing systems, with initial improvements largely realised through a process of trial and error. Handwritten letters from Ambrose Fleming to 'My Dear Major' (Major Flood Page) provide concise summaries of the problems being encountered and possible solutions (and their possible cost!). Fleming's handwritten 'Dear Mr Marconi' letters are similarly to the point and constructive. Earthing was one of the reasons why Goonhilly was dismissed as an option for a larger version of Poldhu – 'Ground too swampy, seems bad for earths' – though 60 years later the installation of Satellite Aerial Number 1 showed that this was a problem that could be overcome.

Putting aerials on ships was also initially a process of trial and error. The Atlantic Leap was presented to Marconi board members and investors as a means to several ends – one end being the end of the subsea cable industry, but the other being revenues from maritime safety, and the provision of news services to transatlantic liners (mostly based on Reuters newsfeeds) and telegram services for passengers. The addition of longwave antennas to ships had to take into account wind loading and stability in rough seas. Cunard was a particularly difficult, though profitable, customer.

Many other issues would seem eerily familiar to contemporary communication network engineers. Writing in 1906, Cuthbert Hall complains about the price (£22!) that the Post Office has quoted for altering the line between Falmouth and Poldhu. Real estate problems and interconnection are an ongoing part of most present-day telecommunication enterprises, including 5G and 6G wireless networks, satellite earth stations and telecom data centres.

Poldhu in these early years was a proving ground for Glace Bay (effectively a mirror image of Poldhu but on the other side of the Atlantic) and Clifden. In the same way that Porthcurno was used as a template for other cable stations, the experimental work at Poldhu helped to shape product development at the Chelmsford factory, which in turn helped to win new overseas contracts.

Later maps of Poldhu show the directional antennas to the south of the original site. Directional longwave at Poldhu was problematic due to space limitations, hence the move to Clifden and the work by Charles Franklin at Poldhu to move to shortwave.

APPENDIX 6:
GOONHILLY EARTH STATION CONSTRUCTION AND RADIO ENGINEERING DOCUMENTS

From contemporary GPO documents (with thanks to Nigel Wall) and notes made by John Bray and the GPO Engineering Department, with additional notes from Husband and Co.

Goonhilly, like Geevor, Porthcurno and Poldhu, was built with state-of the art technology (of its time) and made to work efficiently through the application of good practice engineering. The dishes were – and still are – inspiring examples of electromechanical and radio engineering, from the pouring of concrete through to the design, manufacture and commissioning of a travelling wave tube amplifier (TWTA).

At the heart of the discipline of engineering is the ability to specify requirements with clarity and precision. This appendix provides examples of the art of specification drawn from contemporary documents that do not often see the light of day and are still infusing my office with that indefinable smell of damp musty paper.

Construction and Performance Aerial Number 1
Aerial Number 1 (Arthur): Source John Bray

The Tower was designed by the Ministry of Public Building and Works. The principal building contractors were Peter Lind & Co, and construction began in June 1961. The tower was conceived as a narrow cylinder of glass, steel and concrete. The aerials that would be fixed to the tower could not be allowed to move – or no more than the slightest movement – otherwise there would be significant loss of signal power, so maximum stability was crucial. The base rested on a 90-foot square, 3ft thick concrete raft which was sunk 24ft below ground level. The base rock was Cornish granite.

Wind Deflection Tests, Aerial Number 1, 1965 Husband and Co.

In June 1965 it was decided to investigate the deflections of the bowl periphery when facing the horizon through a range of deflection tests.

Conclusions: the correction required for wind and positional deflection is not likely to exceed plus or minus 0.25 inches.

Specification W6904 for Azimuth Movement Braking Device for Aerial Number 1 Issue 1 1968 Husband and Co.

The aerial achieves motion in azimuth by means of a motor driven reduction gear and a horizontally transposed roller chain which passes around the outside of the azimuth bearing roller track housings at a level slightly above the rollers. A brake is fitted to the D.C. drive motor.

The braking system shall be so arranged that when the aerial is located on a target for instance in auto track mode, the pressure shall be 1200 PSIG [Pounds per square inch gauge- taking into account atmospheric pressure].

The controlling signal will be provided by Messrs Hawker Siddeley Dynamics Limited under a separate contract, as a modification to the Control and Steering Circuits of the aerial.

Completion of the installation and testing should be completed by 27th January.

Aerial Number 3 mandatory performance characteristics

Aerial Number 3 was partly based on the satellite earth station dish installed in Hong Kong and became the baseline for the specification for the Madley dishes which were built for the transfer of the telephone circuits from Goonhilly in 2008. The dishes and radio equipment were sourced from Mitsubishi.

> The third United Kingdom Earth Station was built at Goonhilly Downs, Cornwall, by the British Marconi Company Limited, between August 1970 and July 1972. A summary of the results obtained during the commissioning of the equipment was submitted to the Director of Intelsat Management on 2 August 1972. Following provisional approval for Goonhilly 3 to transmit to the Atlantic INTELSAT IV F2 Satellite, the first carriers to Jamaica and Trinidad were aligned on 5 August 1972 in accordance with the carrier frequency plan AT-OP 8B.

Aerial 3 Post Office Document TD (Technical Department) 13.2.4

The 97-foot (29.56-metre) fully steerable aerial is capable of movement from horizon to zenith in elevation and in full circle within limits in azimuth.

The aerial is designed to operate in wind conditions shown in Table 1

Conditions	Miles per hour (MPH)	Gusting to MPH
Best Tracking Accuracy	35	55
Hold track (brakes off)		80
Parked at Zenith		130
Drive to stow	60	100
Survive in stow		130

Summarised description of Aerial 3 transmit system

The aerial 3 transmit system has been designed for 5 high power amplifiers with a transmit capability for 5 telephone carriers and televisions signals, the original system will only comprise 4 high power amplifiers (HPA) amplifying a total of 3 telephone carriers, plus television.

The earth station (GSO) to transmit frequencies are in the 6 GHz band, which is achieved in the following stages.

1. Modulator: which translates from baseband frequency to intermediate frequency (I.F.). This also provides amplification and F.M. pre-emphasis, as well as energy dispersal activated in low carrier modulation conditions.

2. I.F. equipment: provides cross site cable amplitude equalisation, amplification, group delay correction for earth station equipment and pre-correction for the satellite. Band limiting filters are also included in the I.F. cabinet.

3. Transmit drive equipment: this provides frequency up-conversion from I.F. (70 MHz) to S.H.F. (Super High Frequency 6 GHz) by means of a varactor diode, which acts as a mixer of the I.F. signals and a local oscillator signal; waveguide filters select the upper side band of the mixer output. The drive cabinet being the I.F. to S.H.F. up-converter is the interface between coaxial, narrow band transmission path equipment and microwave (waveguide) equipment.

4. Final amplification before transmission to the aerial takes place in the high power travelling wave tube amplifiers. The H.P.As installed are Marconi P2000 amplifiers, which incorporate two stages of T.W.T amplification, both of which are Siemens valves. This amplifier is a modified version of that used on Goonhilly Aerial 2.

5. The types of transmit carriers to be provided for Aerial 3 when it goes into service are:

Telephony Carrier No.1	252 channel spot beam carrier 6017.5 MHz
Telephony Carrier No.2	Future expansion not yet installed
Telephony Carrier No.3	Future expansion not yet installed
Telephony Carrier No.4	132 channel global beam carrier 6212.5 MHz
Telephony Carrier No.5	132 channel global beam carrier 6118.75 MHz
Television Video Carrier (TVV)	Global Beam Carrier 6403.0 MHz
Television Programme Sound Carrier (TVS)	24 channel beam carrier 6385.75 MHz
Television Cue Sound Carrier (TVQ)	24 channel beam carrier 6383.25 MHz

BIBLIOGRAPHY AND RESOURCES

In addition to the books referenced in the Acknowledgements section (*Cornwall's Communications* by John Moyle, research by David Barlow, and *Lost Sounds* by Alan Renton), the following provide relevant information:

Chapter 1: Tin to telecoms
The Cornish Mining Industry: A Brief History, J.A. Buckley, Tor Mark, Redruth, 2002, reprinted 2008
Industrial Archaeology of Cornwall, A.C. Todd and Peter Laws, David and Charles, 1972
The Official Guide to the Cornwall and West Devon Mining Landscape World Heritage Site published by Cornish Mining World Heritage
The Cornish Miner in America, Arthur Cecil Todd, University of Oklahoma, 1967

Chapters 2 and 3: Steam to smartphones, Broad gauge to broadband
The Men Who Invented Britain, Brian Hannavy, Whittles Publishing, 2024
Brunel, The Man Who Built the World, Steven Brindle, Weidenfeld and Nicolson, 2005
Brunel's Railways: Paddington to Penzance, John Christopher, Bradshaw's Guide, Amberley Publishing, 2013
The Lost Works of Isambard Kingdom Brunel, John Christopher
Isambard Kingdom Brunel, The Engineering Visionary, Eugen Byrne, The History Press, Reissued 2025
The Brunels' Tunnel, Brunel Museum, 2016

Chapters 4 and 5: Subsea cable, and Cable & Wireless
They Talk Along the Deep: A global history of the Valentia Island Telegraph Cables, Donard de Cogan, Radio Society of Great Britain Publications
Lord Kelvin, An Account of his Scientific Life and Work, Andrew Grey, May 2014
Biography of William Thomson, Lord Kelvin, P. Thompson, Sylvanus, 1910
Home Waters, David Bowers, Bloomsbury Publishing, 2023
Changing Places: Porthcurno and the roots of modern communication, PK Porthcurno
A Short History of Cable & Wireless, Cable & Wireless Marine, Chelmsford
The World at Their Fingertips, Geoff Boudreau
Girdle Round the Earth: The story of Cable & Wireless Hugh Barty-King, Heinemann, London, 1979; Adams of Rye, 2018
Voices over the Horizon: Tales from Cable & Wireless, David Souden
Voices of Change: Further Tales from Cable & Wireless, David Souden, Granta Editions, 2001
From Steam to Glass: The curious story of Brazilian Communications, John Packer and Mary Godwin, Cable & Wireless Marine Limited
The Electric Telegraph: A social and economic history, Jeffrey L. Kieve, David and Charles, Newton Abbot, 1973
The Porthcurno Handbook by J.E. Packer, self-published, 1973
'Semaphores to Short Waves', Proceedings of Conference on the Technology and Impact of Early Telecommunications held at the Royal Society for the encouragement of Arts, Manufactures and Commerce (RSA) on Monday 29 July 1996

The Victorian Internet: Tom Standage, Walker Publishing, New York, 1998

Chapter 6: Steam radio to beam radio
Marconi's Atlantic Leap, Gordon Bussey, Marconi Communications, 2000
Marconi: The man who networked the world, Marc Raboy, Oxford University Press, 2016
Guglielmo Marconi, Building the Wireless Age, Tim Wander, New Generation Publishing, 2015
Signor Marconi's Magic Box, Gavin Weightman, Harper Collins, 2004
Wireless over Thirty Years, Richard Vyvyan, George Routledge and Sons, 1933
The History of the Lizard Wireless Telegraph Station, David Barlow
Radio Broadcasting: A history of the air waves, Gordon Bathgate, Radio Society of Great Britain (RSBG)

Chapter 7: The space and satellite and data centre story
The Future of Geography: How power and politics in space will change our world, Tim Marshall, Elliot and Thompson Limited, 2023
From Spark to Satellite, A History of Radio Communication, Stanley Leinwoll, 1979
Then, Now and Tomorrow: The autobiography of a communications engineer, Hardcover John Bray (W.J. Bray), Book Guild Publishing, 25 Nov. 1999
5G and Satellite RF and Optical Integration, Geoff Varrall, Artech House, 2023

Chapter 8: The security story
Tubes: A journey to the centre of the Internet, Andrew Blum, Harper Collins, New York 2012
Listening to the Enemy: St Erth Radio Station 1939–1964, Michael Griffiths, Breton Side Copy, 2022
Behind the Enigma, the authorised history of GCHQ, John Ferris, Bloomsbury Publishing, 2020
The Signalling War: The activities of the Marconi Company during World War One, Elizabeth Bruton
The Evolution of British Sigint 1653–1939, pub. The Stationery Office, Cheltenham, 1997

Additional primary resources directly relevant to the Telecoms Coast

Cornish County archive at Kresen Kernow, Redruth

Cable & Wireless Archive at PK Porthcurno

Marconi Centre, Poldhu

Goonhilly Heritage Centre

If you decide to walk along the Telecoms Coast, you will be following the South West Coast Path www.nationaltrail.co.uk/en_GB/trails/south-west-coast-path/trail-information/

To visit Geevor Tin Mine, go first to www.geevor.com. For the Levant Mine engine houses and Botallack, go to the National Trust website www.nationaltrust.org.uk and the Tin Coast website https://tincoast.co.uk/. The Marconi hut at the Lizard is also owned and managed by the National Trust.

www.nationaltrust.org.uk/visit/cornwall/lizard-point/history-of-the-marconi-radio-stations-on-lizard-point

ABOUT THE AUTHOR

After graduating with a history degree from St John's College, Cambridge, Geoff Varrall worked as a salesman for spot welding equipment then as a product manager for Philips Lighting in Croydon. He joined RTT (Radio Telephony Test Systems) in 1985 as an executive director and shareholder to develop RTT's international business as a provider of technology and business services to the wireless industry. He co-developed RTT's original series of design and facilitation workshops including 'RF Technology', 'Data over Radio', 'Introduction to Mobile Radio', and 'Private Mobile Radio Systems' and he developed the Oxford Programme, a five-day strategic technology and market programme presented annually with the Shosteck Group. Over a 20 year period, more than 5,000 senior level-delegates attended these programmes.

He was a co-author of the *Mobile Radio Servicing Handbook* (Heinemann Butterworth, UK), *Data Over Radio*, (Quantum Publishing, Mendocino, USA) and *3G Handset and Network Design* (John Wiley, New York). Geoff's fourth book, *Making Telecoms Work – from technical innovation to commercial success* (John Wiley) was published in early 2012; his fifth book, *5G Spectrum and Standards*, was published in 2015; his sixth book, *5G and Satellite Spectrum, Standards and Scale*, was published in 2018; and his most recent book, *5G and Satellite RF and Optical Integration*, came out at the beginning of 2023.

Geoff is a faculty member of the Continuing Education Institute, based in Sweden, and presents a five-day workshop on the regulatory issues, competition policy issues, sovereign security challenges and technical and engineering issues of high-count LEO and VLEO satellite constellations including their integration with 5G and 6G terrestrial networks. These workshops include an update on next-generation photonic, optical space and space-to-ground networks and related network operation centre (NOC) capabilities.

As a past Director of Cambridge Wireless, Geoff is involved in a number of wireless heritage initiatives that aim to capture and record past technology and engineering experience. He is a past patron of the Science Museum and a keen supporter of curatorial and archival work on wireless and communication system devices.

ENDNOTES

1. www.goonhilly.org/data-centre
2. https://tincoast.co.uk/
3. www.nationaltrust.org.uk/visit/cornwall/visiting-the-tin-coast
4. https://lizardwireless.org/
5. www.nationaltrust.org.uk/visit/cornwall/lizard-point/lizard-point-southerly-walk
6. https://britishheritage.com/the-life-and-death-of-king-coal-in-south-wales/
7. The Handbook for Travellers in Devon and Cornwall, John Murray 1851 and subsequent editions.
8. www.morsetelegraphclub.com/files/OldSiteArchive/library/files/pdf/TTEH/TTEHPreface.pdf
9. https://www.silicon.co.uk/networks/broadband/bt-recoups-105-million-by-recycling-copper-cables-582346
10. www.visualcapitalist.com/visualizing-the-critical-metals-in-a-smartphone/
11. www.visualcapitalist.com/sp/which-countries-dominate-the-supply-chain-for-strategic-metals/
12. www.nationalwealthfund.org.uk/news/national-wealth-fund-drives-growth-ps286m-investment-cornish-metals-inc-facilitating-domestic
13. The Levant mine is better known to many as Tressiders Rolling Mill in the Poldark TV series. www.cornwalllive.com/news/cornwall-news/new-proof-cornish-been-mining-9341015
14. https://nuttersworld.com/ancient-trade-routes-mediterranean-sea/tin-roads/ The Cornish mining Industry, Buckley, J.A., 1992
15. Arthur Todd, The Cornish Miner in America
16. https://trevithicksociety.info/ www.nationaltrust.org.uk/visit/cornwall/levant-mine-and-beam-engine
17. https://www.countrylife.co.uk/property/a-glorious-clifftop-home-built-by-the-man-who-escaped-the-cornish-tin-mines-to-become-an-international-diamond-trader-267503
18. Subsea cables were operated from banks of batteries. Once dynamos were discovered the default was to use lead acid batteries keeping them charged or recharged on a duty cycle. See Appendix 3 for more detail on this.
19. www.atlanticcouncil.org/blogs/energysource/is-power-ever-too-cheap-to-meter/
20. www.nationalgrid.co.uk/innovation/projects/isles-of-scilly
21. https://globalmarine.co.uk/
22. www.bbc.co.uk/news/uk-england-cornwall-39166226
23. www.viking-link.com
24. https://celticseapower.co.uk/realising-cornwalls-floating-offshore-wind-ambitions/
25. https://electrichome.uk/news/national-grids-31bn-project-to-install-thousands-of-high-voltage-pylons-across-britain
26. In the 1980s in Luxulyan, a village on Bodmin Moor close to what is left of Marconi's high-power shortwave radio station, locals and people from around Cornwall staged a six-month occupation of farmland to prevent test drilling by the Central Electricity Generating Board investigating the area as a potential nuclear power station site.
27. https://newatlas.com/energy/france-tokamak-cea-west-fusion-reactor-record-plasma-duration/
28. www.home.sandvik/en/stories/themes/electrifying-the-future/
29. www.newscientist.com/article/2290944-how-electrification-is-changing-mining
30. https://celticseapower.co.uk/realising-cornwalls-floating-offshore-wind-ambitions/
31. A series of travel guidebooks was published in London by John Murray, beginning in 1836.
32. 30 million bricks were used to line the Box Tunnel; a ton of candles and a ton of gunpowder were being used every week and by the time the tunnel was completed about 100 workers had been killed.
33. The *Great Eastern* was six times larger than any other ship, with a 1,000-horsepower steam engine driving the paddles, a 1,600-horsepower engine powering the propeller, 72 furnaces to produce steam and 65,000 square yards of sail. Shipping coal from Welsh coal pits to distant parts of the Empire was expensive. The *Great Eastern* could load 18,000 tons of coal.

ENDNOTES

34. Daniel Gooch was later to become the Chief Engineer for Mr Pender's new subsea manufacturing and contracting business, the Telegraph Construction and Maintenance Company (Telcon).
35. Mr Marconi's route to Poldhu, www.helstonrailway.co.uk
36. The Corn Laws marked an important step towards free trade, but in the short term meant that corn could be imported without tariffs being applied, reducing the realised prices for Cornish farmers.
37. GWR replaced steam engines with diesel hydraulic rather than diesel electric engines. Diesel hydraulic engines have inlet and exhaust valves that are hydraulically operated. The engines are quieter and have lower maintenance than mechanical systems (though mechanical systems are theoretically more efficient if well maintained). GWR introduced Hitachi bi-mode trains in 2018 capable of running on overhead electric track or in diesel mode and is now testing tri-mode trains, engines with batteries for their next generation locomotives. www.railadvent.co.uk/2025/07/great-western-railway-publishes-fast-charge-battery-findings-following-success-of-year-long-trial.
38. Dan Cleaver, *History of Porthcurno*. As an eight-year-old in 1870, Cleaver had watched the first cable come ashore on Porthcurno beach.
39. John Liffen, Curator of Communications at the Science Museum in London, 'The Definitive History of Cooke and Wheatstone's Earliest Telegraph Instruments and their Use between 1837 and 1842' July 2010. The 'Making the Modern World' and the 'Information Age' galleries at the Science Museum mark Cooke and Wheatstone's contribution to both the railway industry and telecommunications in the 19th century.
40. One of John Pender's friends and new part-owner of the SS *Great Eastern* was Daniel Gooch, an MP and previously locomotive superintendent of the GWR. The investors paid $125,000 for the *Great Eastern*, which had cost $5 million to build. Funnel Number 4 and two boilers were removed to make room for three cable tanks containing the 2,600 miles of cable weighing 9,000 tons. As mentioned in Chapter 2, Daniel was later to become the chief engineer for Mr Pender's new subsea manufacturing and contacting business, Telcon.
41. https://www.britannica.com/biography/Charles-Tilston-Bright. At 26, Charles Tilston Bright was one of the youngest people ever to be awarded a knighthood.
42. John Packer in *Gateway to Empire* (RSA Semaphore to Short Waves, 1998) makes the point that an ill-fated attempt to establish a ship-to-shore cable service in Porthcurno Cove had left the legacy of an overhead pole route to Penzance, which was an added advantage.
43. https://atlantic-cable.com/Article/GuttaPercha/index.htm
44. See also Appendix 3. Accumulators produced hydrogen and oxygen at the end of their recharge cycle and were therefore an explosion risk.
45. Given that the US stock market crashed in 1929, the merger between ETC and the Marconi Company, which had been orchestrated in the 1928 Conference largely at the instigation of the government (and accepted with some reluctance by the ETC Board and Marconi), was fortuitous.
46. 'Thunder factory' was a disparaging term used by Neil Maskelyne, employed by the Eastern Telegraph Company at Porthcurno to listen to Marconi transmissions from Poldhu.
47. This is the link for www.cornishlithium.com
48. Morse code could only be used slowly on subsea cable due to the time-dispersion effects of the cable and surrounding sea water.
49. A nautical mile is defined as the meridian arc length corresponding to one minute/one sixtieth of a degree of latitude at the equator. A fathom comes from the old English 'faethm' meaning 'outstretched arms'. It was used for nautical soundings from 1600 onwards, with a knot being made for each fathom-length of rope. There are 1,013 fathoms in a nautical mile.
50. The telegraph plateau turned out to be a myth based on inaccurate measurement; the Appendix 4 case study on Valentia Island covers this in more detail.
51. In 1902 the pass standard for operators trained in Porthcurno was one error every 1,000 words. The bit error rate in a modern optical fibre cable is typically specified as 1 bit error for every 100,000,000,000 bits (1 in 10^{10} bit error rate).
52. Rowena Cade, who built the Minack Theatre on the cliffs above Porthcurno before and after the Second World War, regularly hosted performances of plays by Shakespeare. The theatre remains busy, selling 80,000 tickets each year for an eclectic mix of theatre and musical events.

53 The Second Boer War, from 11 October 1899 to May 1902, was expensive for Britain, costing £211 million – well over $20 billion at 2025 prices – with a casualty count higher than the Crimean War (1853–1856).
54 Telecoms engineering in the Second World War was a protected profession and was applied to engineers working on integrated landline and radar systems.
55 At the outbreak of the war, the ETC had nine cable ships and the Post Office had several more. Electra House at 84 Moorgate in London, a Grade II listed building dating from 1903, became part of the Second World War Special Operations Executive known as Department EH-Electra rather than Elettra (see Marconi) House. 'Electra' and 'Elettra' are both derived from the Greek 'elektra' for amber, which was how the Ancient Greeks discovered (static) electricity.
56 See Hugh Barty-King, *Girdle Round the Earth*, for additional detail on Cable & Wireless up to the 1970s.
57 See Television Broadcast History, the Roving Eye, 2020, Broadcast Engineering Conservation Group, compiled by Richard Harris.
58 The early systems pioneered by the American Bell Company used two hard-drawn copper #12 wires. These cables successfully spanned distances of over 200 miles
59 The first manual (plug and socket) exchange was installed in London in 1878, but Britain was still behind the US in terms of voice telephony deployment.
60 By the 1950s the concept of link budgets across all communication links including radio was ubiquitous and simply involved taking the power applied to the cable or wireless system in decibels and then subtracting whatever losses (including noise and distortion) were introduced by the transmission medium. The decibel is one tenth of a bel. derived from Alexander Graham Bell's way of describing the very large differences in signal level in acoustic and radio and wireline systems using small numbers. Simply put, a decibel expresses the ratio of two values of transmitted or received power on a logarithmic scale, and really that is all we need to know.
61 At the height of the Cold War in the late 1960s, there was a growing recognition that telecommunication switching protocols were vulnerable to disruption. What was needed was an address protocol that could adaptively deliver traffic across multiple routes. This became ARPANET (the US Advanced Research Projects Agency Network). These protocols evolved in the 1970s to become the Internet. The Internet allowed a new generation of companies (Google, Amazon, Meta Group, Apple and fellow travellers) to provide Over the Top (OTT) Services to end users over networks that, due to competition, were forced to provide bandwidth at below the cost of delivery. The value of the OTT companies increased exponentially. Network provider stocks values headed in the opposite direction.
62 A receiver station was also built at Letterfrack, further down the west coast of Ireland, with eastbound messages coming from Marconi Towers, the high-power wireless station in Nova Scotia.
63 More information on the gilded age is here https://www.goodreads.com/book/show/98906.The_Gilded_Age
64 Hertz died on 1 January 1894, aged 37, possibly of bone cancer and/or complications from surgery.
65 In 1864 James Clerk Maxwell (1831–1979) published his *Dynamical Theory of the Electromagnetic Field*, based on four equations that express the properties of electric fields, magnetic fields, electric charge and electric current.
66 Annie Jameson was to take a close interest in Guglielmo's career and general wellbeing all her life – and she lived into her eighties. As a result, she had occasionally tense standoffs with Beatrice, Marconi's wife. Mary Elizabeth Horsley, Isambard Kingdom's wife and a doyenne of the London social scene, had a similarly prickly relationship with Isambard Kingdom's mother, Sophia.
67 A modest amount of directivity was achieved at Clifden with an inverted L-shaped antenna lying on its back, the long bit of the L pointing towards America. See Clifden in main text.
68 *Wireless over Thirty Years*, Richard Vyvyan, George Routledge and Sons, 1933.
69 The letter S was used because it inflicted the minimum wear and tear on the spark discharger.
70 The difference between day and night was partly due to a higher noise floor during the day (heat is noise, so radio noise increases when the sun shines) but also because the D layer – the lowest layer in the ionosphere (at 60 kilometres) - becomes charged with solar energy during the day and, because it is denser, attenuates the longwave transmissions, preventing

the radio waves reaching the upper layers (E at 120 kilometres and F at 150 kilometres) which is what the long waves bounce off at night.

71   The capacitance of the kite-flown aerial used in Newfoundland would have constantly changed, which in turn would have affected the matching of the small amounts of receive energy being heard in Marconi's headphones. This is now known as 'noise matching on the receive path'. On the transmit path in modern radio systems, matching efficiency (specifically, power-matching efficiency) is measured by the term 'voltage standing wave ratio' (VSWR). A high VSWR is undesirable, as reflected energy (energy not being coupled efficiently into the antenna) arrives back at the output stages of the transmitter power amplifier.

72   For his grand demonstration to the guests from the International Telegraph Conference visiting Poldhu, it was probably not a coincidence that Marconi chose the intermediate landing stations for the subsea cables from Porthcurno as the destination points for his Marconigram.

73   Newquay is a horizontal spaceport; rockets are taken up to 50,000 feet under the wings of aeroplanes and shot into space. The advantage is that satellites can be launched in any direction including what is called prograde, when rockets take off in the direction of the Earth's rotation (which gives them a flying start), or retrograde, when they take off the other way. Sun synchronous and dawn-and-dusk synchronous satellites are a special case of retrograde orbits. Richard Branson made a brave attempt to get Newquay up and running using a converted Boeing 747. This large plane has a fifth mounting point under the wing to allow it to carry a spare engine around the world for maintenance and repair. This is a useful place to put a rocket launcher. Unfortunately, the business failed to take off, but the runway is still there. (Anorak fact: Newquay was designated as an emergency landing runway for Concorde in case it couldn't make it back to Heathrow when flying from New York.)

74   Geostationary satellites are increasingly used as relays to Low Earth and Medium Earth Orbit satellites using RF and optical transceivers.

75   The class of orbits (defined by altitude rather than inclination) known as Medium Earth Orbits are defined as anything higher than 2,000 km (1,243 miles, the upper limit defined for Low Earth Orbits) and lower than Geostationary Orbits at 35,786 km (22,236 miles) above sea level. The most common and important satellites in Medium Earth Orbit are the GPS satellites at an altitude of 20,300 km (12,550 miles). An additional class of satellites is described as geosynchronous. As with geostationary satellites, they are travelling at the same time and in the same direction as the Earth (anticlockwise) but are not positioned directly above the Equator. This means that from an earth station's perspective, they travel in a figure of eight, which means the satellite dish has to follow the satellite. They are useful for covering long, thin north-to-south countries such as Japan.

76   Goonhilly is at 50.1 degrees north, 5.3 degrees west, which means it can see what is called the geostationary arc from 65 degrees east through 0 degrees to 75 degrees west with clear horizons in all directions down to below the 5-degree elevation.

77   Telstar 1 stopped working after sending 400 telephone, telegraph, fax and TV transmissions; it temporarily recovered, but then on 21 February 1963 failed completely. The Telstar name was still used to name satellites through to 2004.

78   Laying cables under roads is often a good option for terrestrial cables, as it involves relatively few government and local agencies (typically the local council and Department of Transport). You can trace cable routes by driving along and looking at manhole covers.

79   The Post Office Engineering Department based at 207 Old Street is close to where Inmarsat and Viasat are now headquartered.

80   *The Autobiography of a Communications Engineer*, by John Bray, Book Guild, London, 1999, has additional background on the history of the GPO.

81   The lines from each station in the Chain Home system needed to be electrically the same length, so that signals from each station stayed in synchronisation with each other. The whole network was synchronised to the AC mains, based on 25 pulses per second coupled to 50 Hz. This puzzled the Germans, who had flown a Zeppelin slowly along the south coast to try and work out how the radar system worked. The principles were similar to Marconi's four-tower direction-finding system; this had been developed by Bellini and Tosi, who had become part of the Marconi Company in 1912. Captain Round did some work on the system in Poldhu. Marconi's radio direction finding systems were a type of goniometer (from the Greek 'gonia' meaning angle and 'metron' meaning 'to measure'). Skilled operators could resolve signal angles in azimuth and elevation derived from the phase difference

	between the four towers (and top and bottom of the towers). Elizabeth Bruton has done scholarly research on radio signalling and direction finding in the First World War (see Bibliography).
82	The Early Bird Intelsat satellite and subsequent TV and telephony satellites were (and are) in Geostationary Orbit.
83	The frequencies supported at Goonhilly at time of writing are VHF (for legacy telecommand and telecontrol), UHF, L Band, C Band, X Band (for military satellite systems) and Ku and Ka band (TV and LEO).
84	Hubble has now been joined in space by the James Webb telescope. This $30 billion telescope hovers at what is called the L2 Lagrange point, which is one of five points where the Earth's gravity is balanced by the gravitational pull of the Sun.
85	An exabyte is 1,000 petabytes.
86	The new generation of earth stations being installed today includes a Network Operations Centre (NOC) which controls the space constellation. This is different from the Network Operation Centre for a GSO Earth Station, where the satellites more or less stay in the same place. The LEO satellites are highly autonomous – for instance, they are capable of doing their own collision avoidance – but overall control of the constellation is still done from Earth. Machine Learning (ML) is the process by which RF and optical link performance between satellite nodes in space is analysed to optimise space network link architecture.
87	Military grade 5G base stations are now being launched into Low Earth Orbit highlighting how new satellite constellations are introducing new sovereign security challenges. See Lockheed Martin 5G.MIL Unified Network Solutions.
88	Not everyone makes it into the history books, and there are thousands of engineers that worked on security systems or radar systems in the Second World War who then went on to help build some of the most successful radio and telecommunication companies in the UK. My dad, Jack Varrall, (born in a Zeppelin air raid in 1917) worked at Dollis Hill in the Second World War and then joined British Control and Communications (BCC) one of the many British-owned companies that prospered in the 1950s. He became Managing Director of Airmec in High Wycombe, designing and manufacturing signal generators, at least a few of which would have found their way into radio stations along the Telecoms Coast and can occasionally be found in radio museums, including RAF museums with radio heritage collections.
89	https://brycetech.com/reports/report-documents/Bryce_Launch_Sites_2025.pdf
90	The Five Eyes are the United States, the United Kingdom, New Zealand, Australia and Canada, known colloquially as the Anglosphere intelligence alliance.
91	Augustus De Morgan (1860–1871) was a mathematician who deserves more recognition. He established the theory of limits and boundaries, which is widely used in AI search algorithms.
92	PK Porthcurno collaborated with GCHQ in 2022 on an exhibition based on images, artefacts and film footage from the GCHQ archive called 'Watcher of the Skies'. The exhibition, curated by David Twomlow, remained open until June 2023.
93	The story of a disconnected Tonga is told here: www.datacenterdynamics.com/en/news/tongas-only-domestic-subsea-suffers-another-outage-could-take-weeks-to-repair/
94	In 1874 Siemens Brothers – the London-based rather than German-based bit of the company – launched a new cable ship, the *Faraday*. This boat had twin screws, a bow rudder and a twin superstructure with a through deck, designed to be equally efficient at laying or picking up a cable. Carl Siemens was the engineer in charge.

INDEX

Anderson, Sir James   46
Bayes, the Reverend Thomas   128
Becquerel, Edmund   24
Bell, Alexander Graham   4, 78, 106
Bennett, Gordon   94
Berryman, Lieutenant O.H   61
Bessemer, Henry   3
Bezzi Scali, Christina   87
Boole, George   128
Bray, William, John   117-118, 120, 154-155, 160, 168
Bright, Charles Tilston   45, 65, 141, 144, 165
Brunel, Isambard Kingdom and family   12, 30-36, 38-43
Cade, Rowena (Minack Theatre)   165
Caesar, Julius   11, 12, 41
Churchill, Winston   75
Clarke, Arthur C   14
Cooke, William   42-43, 48, 58-59, 165
Daniell, John Frederic   42, 57
Davis, Henry Jameson   92, 98, 148
Dayman, Joseph   62
De Morgan, Augustus   128, 169
Dee, John   124
Denison, Emma   44
Edison, Thomas   6, 25, 57, 89, 94, 97, 105
Einstein, Albert   127
Faraday, Michael   vi, viii, 19, 57, 89, 94, 96, 97, 103, 107, 119, 133-134, 138, 148-149, 152
Field, Cyrus   143-144
Fleming, Ambrose   vi, xi, 6, 20, 53, 56, 89-90, 94-97, 103, 107, 119, 133-134, 138, 148-149, 152
Flood Page, Major   xiii, 98, 101, 148, 150, 152
Fourier, Joseph   145
Franklin, Benjamin   56
Franklin, Charles   89, 94-96, 106-109, 115, 134, 140, 153
Gooch, Daniel   36, 164, 165
Hall, Cuthbert   148, 150, 152
Heaviside, Oliver   4
Hertz, Heinrich,   91, 115, 141, 167
Hill, Rowland   54
Huntsman, Benjamin   54
James, Mrs M.C   55, 75
Jameson, Annie   xv, 68, 91, 92, 167
Jansky, Karl   117
Joule, James Prescott   17
Kemp, George   20, 96, 98, 103, 105, 119, 148, 150
Kruger, Paul   70
Lipton, Sir Thomas   90

Lodge, Oliver   vi, 93, 141
Macadam, John   41
Marconi, Gugliemo (and company)   20, 27, 34, 37, 41, 45, 49, 51, 53, 59, 68-71, 73-79, 85-96, 98-113, 115-116, 119, 125, 140, 144, 148-152, 156, 160, 164-169
Maskelyne, Neville   51, 57, 125, 165
Maudsley, Henry   31
Maury, Matthew Fontaine   144
Maxwell, James Clark   91, 94, 141, 167
Morse, Samuel   xi, 58, 63-66, 102, 163-165
Moyle, John   v, 63, 158
Murray, John   30, 163-164
Musk, Elon   63
Mussolini, Benito   90-91, 111
Newcomen, Thomas   16, 40, 159
Newton, Isaac   14
O'Brien, Beatrice   87, 99, 107, 167
Oats, Francis   17, 18, 19
Packer, John   xii, 75, 77, 141-142, 159, 165
Paget, Percy   96, 105, 148
Pender, John (and the other Penders)   xii, xv, 8, 10, 35, 44-49, 53, 61, 63, 67, 68, 74, 76, 90, 131, 143, 149, 164-165
Perkins, Veronica Davis   xv
Portillo, Michael   38
Preece, William   xv, 65, 78, 88, 92, 117
Renton, Alan   vii, 158
Rhodes, Cecil   10, 18, 69-70
Rickard, James   11
Righi, Augusto   91
Rolls, Charles and Royce, Henry   27, 28
Round, Henry, Captain   vii, 100
Savery, Thomas   16
Siemens, William (and company)   19-20, 52, 57, 62, 74, 157, 170
Snowden, Edward   131-132
Stephenson, Robert   36
Strutt, John William (Lord Rayleigh)   viii
Telford, Thomas   41
Tesla, Nikola   25
Thomson, William, (Lord Kelvin)   143-146, 159
Trevithick, Richard   17, 32, 40, 163
Turing, Alan   126
Volta, Allesandro   56
Vyvyan, Richard   20, 53, 96-99, 103, 133, 148-149, 159, 167
Walsingham, Francis   124
Watt, James   16, 17, 40, 57
Westinghouse, George   25, 149
Wheatstone, Charles   42-43, 48, 57-58, 60, 62, 126, 165
Whitworth, Joseph   62